D0473640

To my little boy, Draco, who isn't so little anymore! You continue
to be my muse, in so many ways. You are the best parts of Mr. Mouse,
and little Piddles is my Ms. Squirrel. . . . I love you!

Copyright © 2020 by Danica McKellar
Front cover design by Mike Verta

All rights reserved. Published in the United States by Crown Books for Young Readers,
an imprint of Random House Children's Books, a division of Penguin Random House LLC, New York.

Crown and the colophon are registered trademarks of Penguin Random House LLC.

Visit us on the Web! rhcbooks.com

Educators and librarians, for a variety of teaching tools, visit us at RHTeachersLibrarians.com

Library of Congress Cataloging-in-Publication Data is available upon request.
ISBN 978-1-101-93402-9 (trade) — ISBN 978-1-101-93403-6 (lib. bdg.) — ISBN 978-1-101-93404-3 (ebook)

The text of this book is set in 12-point Bliss.

MANUFACTURED IN CHINA
10 9 8 7 6 5 4 3 2 1
First Edition

THE TIMES MACHINE!

LEARN MULTIPLICATION AND DIVISION . . . LIKE, YESTERDAY!

DANICA McKELLAR

ILLUSTRATIONS BY
JOSÉE MASSE

Crown Books for Young Readers ♛ New York

"New Math" vs. Memorizing Multiplication Facts!

A Note to Parents

In this book, I'm thrilled to present some super fun and easy ways to learn the multiplication facts—something I wish I had growing up! More on those in a moment . . .

As kids, many of us learned our multiplication facts—times tables—by tiresome memorization and drills, and then applied those (painfully attained!) memorized facts to more complicated multiplication and division problems. Phew! It was often boring and more labor-intensive than we liked, but it worked, right?

Nowadays, the "new math" takes a much more visual approach, focused on teaching what it *means* to multiply and divide. Everything gets broken down into many more steps. While this can be useful in some ways, often it leads to overly complicated-looking third and fourth grade homework and parents wanting to pull their hair out! (See my "New Math" Translation Guide—for Grown-Ups! at the back, and save that gorgeous hair of yours.)

The other issue is that teachers often spend so much of their lessons covering these new teaching methods that they don't have enough time to help kids memorize their multiplication facts—the foundation upon which much of mathematics is built.

In this book, I take a two-pronged strategy. Yes, I cover the newfangled ways multiplication and division are taught—even for multi-digit multiplication and long division—to make sure kids have success on tests, and to help parents figure out what the heck is going on with the foreign-looking homework! But what I'm equally excited to share is Chapter 5: The Core of the Times Machine. This chapter is filled with truly entertaining ways for kids to learn their multiplication facts, once and for all. You'll find stories, poems, and fun pictures to help kids learn all of their facts—which will make a huge difference in their success. Being fluent with the times tables (multiplication facts) is key to being able to tackle fourth and fifth grade math with ease—and the rest of middle school and high school math, for that matter!

Congratulations, and have fun experiencing this book with your child!

Danica

P.S. Most kids in third and fourth grade won't have any trouble reading this book, but I encourage you to enjoy some of it together! You might find yourself giggling at the antics of Mr. Mouse and Ms. Squirrel. . . .

Throughout the book are "Game Time" sections with practice problems. You'll notice that I didn't leave room in the book for your child to do the math, so you'll want to have a separate piece of paper handy. This was a deliberate choice to keep the book pages clean so your child can do the problem sets more than once for better mastery of the material. As an added bonus, siblings or friends will be able to use the book as well. Have fun!

Introduction

RIGHT. ANYWAY, FOLLOW ME, GUYS!

THIS IS THE TIMES MACHINE, A SECRET MAGICAL DEVICE JUST FOR KIDS READY TO LEARN THEIR MULTIPLICATION FACTS--YOU KNOW, THEIR *TIMES* TABLES!

WAIT--IS THIS A *TIME MACHINE*?!

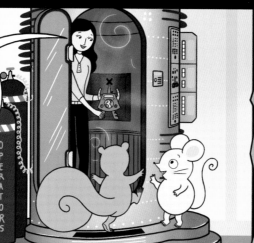

TIMES MACHINE. ONCE WE STEP INSIDE, THIS REMOTE CAN SEND US ANYWHERE AND ANY*TIME* IN HISTORY TO HELP US LEARN ABOUT MULTIPLICATION AND DIVISION. WE CAN CHOOSE WHERE WE GO, OR IT CAN CHOOSE FOR US!

TIMES MACHINE, TAKE US TO MY LIVING ROOM!

POOF!

YOUR LIVING ROOM? THERE'S NO SUCH THING AS TIME TRAVEL. I KNEW IT. WAS THAT A TRAPDOOR OR SOMETHING?

HAVE PATIENCE, MR. MOUSE . . . MY HOUSE IS JUST A BETTER PLACE TO TRAVEL TO AND FROM, AND WE'LL START EACH CHAPTER FROM HERE.

OOH, THE MAGIC OF MATH-- I'M SO EXCITED TO START LEARNING!

FIRST WE'LL TALK ABOUT WHAT IT MEANS TO MULTIPLY TWO NUMBERS, WE'LL LEARN SOME TRICKS TO MEMORIZE THE FACTS-- AND YES, WE'LL EVEN LEARN A LITTLE HISTORY ALONG THE WAY. LET'S DO IT!

Oh, the Places We'll Go . . . in the Times Machine!

Chapter 1

Magic Marching Ants and Ancient Rome:
Intro to Multiplication

Marching Ants: Arrays for Multiplication!

Let's say there's a group of ants marching, and we want to know how many there are. We see 4 rows and 5 columns:

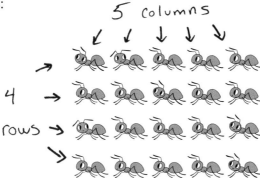

Let's group together each row, so that we have 4 groups of 5 each:

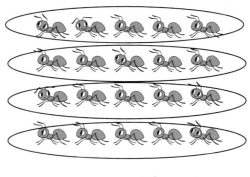

4 groups, with 5 in each group

Hmm, how can we tell how many *total ants* there are? Well, we could count them all, but that would take a while. What if we add them up, row by row? So that would be 5 + 5 + 5 + 5. We could start with 5 in the first row. Then 5 + 5 = 10 for the first two rows, then add in the third row, 10 + 5 = 15, and finally, for the fourth row of 5, we have 15 + 5 = **20**. Phew! That was a lot of work, and we had to be careful to add 5 the correct number of times. If we wanted to be fancy, we could figure out how many ants there are by *skip counting* by 5's four times: 5, 10, 15, **20**. (See p. 32 for more on this.)

Or we could just "know" that 5 × 4 = 20.

Multiplication is repeated addition. For example:

★ 3 × 2: "three times two" means "three, *two times*"; in other words:

$$3 + 3$$

★ 6 × 4: "six times four" means "six, *four times*"; in other words:

$$6 + 6 + 6 + 6$$

★ 9 × 8: "nine times eight" means "nine, *eight times*"; in other words:

$$9 + 9 + 9 + 9 + 9 + 9 + 9 + 9$$

But before we learn our multiplication facts, we'll spend a few chapters learning how to *think* about multiplication. A great way to see what multiplication looks like is to arrange objects in rows and columns, like the ants were. These are called **arrays**.

An **array** is an arrangement of objects in rows and columns. For example:

3 columns

4 rows

★ ★ ★
★ ★ ★
★ ★ ★
★ ★ ★

This array has 4 rows and 3 columns. It could be called a "4 by 3 array."

In any array, if we count the rows and columns, we can write down the number sentence the array is showing us. Below, we see 2 rows and 5 columns, so the multiplication problem is 2 × 5—there are 2 groups of 5 each, after all!

● ● ● ● ●
● ● ● ● ●

$$2 \times 5 = 10$$

↑ rows   ↑ columns   ↑ total number of objects

There are 10 dots total, and the number sentence that describes this picture is 2 × 5 = 10. And our "stars" array above describes the multiplication sentence 4 × 3 = 12. Not so bad, right?

Ancient Rome: Rows vs. Columns

Rows and Columns

Usually in a multiplication problem, the number of rows comes first. So for 4 × 5, the array would have 4 rows and 5 columns. To remember that rows come first, I like to imagine ancient Romans *rowing* across the sea to go see some really tall *columns*.

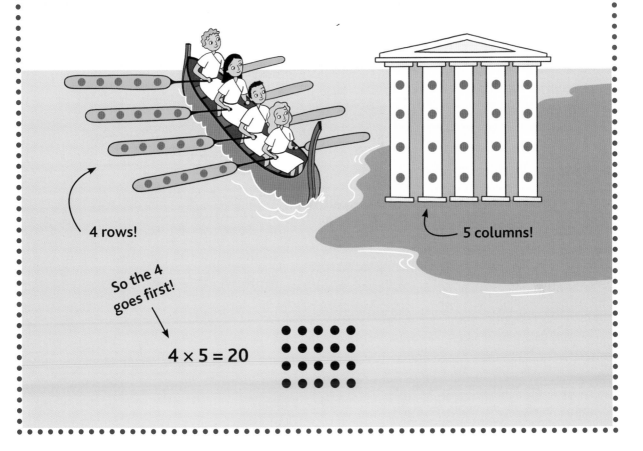

4 rows!

5 columns!

So the 4 goes first!

4 × 5 = 20

Let's practice what we know about multiplication arrays!

Answer the questions, and then figure out what multiplication problem the array is showing us. I'll do the first one for you!

1.

How many rows does this array have? How many columns?

__?__ × 6 = 18

Let's Play: Hmm, which are the rows, and which are the columns? Let's imagine those Romans rowing out to sea, and imagine drawing the oars on top of the array. We see that there are 3 rows!

Next, we can imagine drawing some tall Roman columns over the array instead, and we see 6 columns. So the multiplication problem must be 3 × 6, which means the missing number is 3!

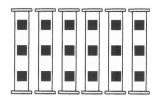

Answer: 3 rows, 6 columns; 3 × 6 = 18

2.

How many rows?

How many columns?

2 × __?__ = 16

3.

How many rows?

How many columns?

3 × __?__ = 9

4.

How many rows?

How many columns?

__?__ × 4 = 12

5.

How many rows?

How many columns?

$4 \times \underline{\ ?\ } = 12$

6.

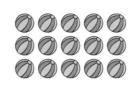

How many rows?

How many columns?

$3 \times \underline{\ ?\ } = 15$

7.

How many rows?

How many columns?

$\underline{\ ?\ } \times 6 = 24$

8.

How many rows?

How many columns?

$5 \times \underline{\ ?\ } = 25$

9.

How many rows?

How many columns?

$\underline{\ ?\ } \times 7 = 21$

10.

How many rows?

How many columns?

$4 \times \underline{\ ?\ } = 36$

(Answers on page 220.)

A Trip to the Toy Factory: Factors and Products

Just like **factories** <u>make</u> **products**, two **factors** <u>make</u> a **product**. And that's a great way to remember the parts of any multiplication problem!

Factors and Products

In any multiplication problem, the two numbers being multiplied together are called the **factors,** and the answer is called the **product.** For example, in 3 × 5 = 15, the two factors are 3 & 5, and the product—the thing they make—is 15. And the factors 10 & 7 multiply to make the product 70.

$$3 \times 5 = 15 \qquad 10 \times 7 = 70$$

factor   factor   product   factor   factor   product

SO, WAIT, THE FIRST *FACTOR* IN A MULTIPLICATION PROBLEM IS THE NUMBER OF ROWS?

IF WE ARE MAKING ARRAYS, YES, THAT'S EXACTLY RIGHT. CHECK IT OUT.

28 dots total!

4 rows

7 columns

factors   product

$$4 \times 7 = 28$$

The Commutative Property

In multiplication, *the order of the factors* doesn't change the product. In other words, 2 × 5 gives the same answer as 5 × 2. They both equal 10! After all, look what happens if we push this array on its side:

$5 \times 2 = 10$

↑   ↑   ↑

rows   columns   total number of objects

Push!

$2 \times 5 = 10$

↑   ↑   ↑

rows   columns   total number of objects

See? We've switched the rows and columns—the factors *changed places*. But we haven't changed the *number* of dots, so the answer is the same! (For more, see Chapter 6.)

HEY, THAT MEANS IF WE KNOW A MULTIPLICATION FACT LIKE 4 X 7 = 28, WE ALSO KNOW THAT 7 X 4 = 28. THERE'S LESS TO MEMORIZE THAT WAY!

YES! *THAT'S* WHAT I'M TALKING ABOUT!

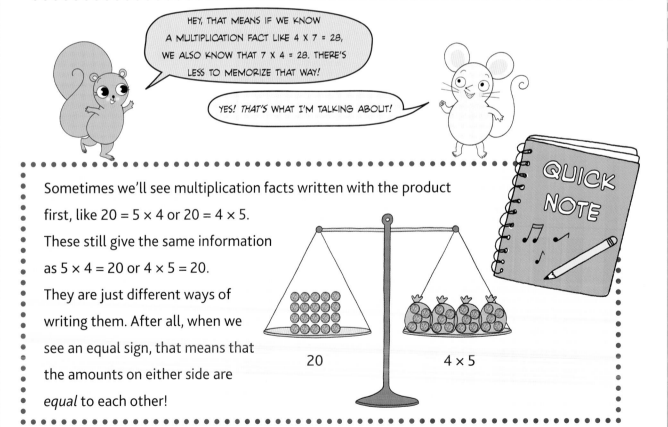

Sometimes we'll see multiplication facts written with the product first, like 20 = 5 × 4 or 20 = 4 × 5. These still give the same information as 5 × 4 = 20 or 4 × 5 = 20. They are just different ways of writing them. After all, when we see an equal sign, that means that the amounts on either side are *equal* to each other!

20   4 × 5

GAME TIME!

For each math sentence, name the *factors* and the *product*, and then switch the order of the factors to name another multiplication fact. I'll do the first one for you!

1. 72 = 8 × 9

Let's Play: Hmm, let's not be confused by the order! No matter what comes first, the *factors* are the numbers that are multiplied times each other, so that's the 8 and 9, right? The *product* is the result of that multiplication—the 72! This math sentence, 72 = 8 × 9, is really the same as 72 = 9 × 8, and also 8 × 9 = 72 and 9 × 8 = 72. But the problem just asked us to switch the order of the factors (the 8 and 9), so that's the math fact we'll put in our answer. Done!

Answer: 8 and 9 are the factors; 72 is the product. 72 = 9 × 8.

2. 4 × 6 = 24

3. 5 × 3 = 15

4. 24 = 4 × 6

5. 6 × 7 = 42

6. 56 = 7 × 8

7. 0 × 7 = 0

8. 20 = 10 × 2

9. 8 × 6 = 48

10. 7 × 9 = 63

11. 21 = 3 × 7

12. 30 = 5 × 6

13. 1 × 3 = 3

(Answers on page 220.)

Many Mice: Multiples

Having multiple toys means having more than one *toy*. Having multiple 3's means having more than one *3*. So if we have two 3's, that's 6. If we have three 3's, that's 9. In other words, 3 × 2 = 6, and 3 × 3 = 9. We're just doing a little multiplication!

The **multiples** of any number are what we get when we multiply it times the *counting numbers,* like 1, 2, 3, 4, 5, etc. For example, some multiples of 3 are 3, 6, 9, 12, and 15. This is because:

$$3 \times 1 = 3, \quad 3 \times 2 = 6, \quad 3 \times 3 = 9, \quad 3 \times 4 = 12, \text{ and } 3 \times 5 = 15$$

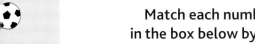

GAME TIME!

Match each number with the group of its first few multiples
in the box below by writing its letter. I'll do the first one for you!

1. 7

Let's Play: Hmm, multiples of 7? Since we haven't learned our multiplication facts, how do we know what its multiples are? Well, we know that 7 × 2 = 14, because that's just two 7's. In other words, it's 7 + 7, which is 14. Then we could add 7 again to get 14 + 7 = 21, which must be the same as 7 × 3. Adding one more 7, we get 21 + 7 = 28. And look, "C" has those same numbers!

Answer: C

2. 10   3. 3   4. 4   5. 12

Multiples

A. 20, 30, 40   B. 8, 12, 16   C. 14, 21, 28

D. 6, 9, 12   E. 24, 36, 48

(Answers on page 220.)

Comparing Numbers—with Multiplication!

Another way to think about multiplication is that we are *comparing* numbers.

By saying 18 = 6 × 3, we're really saying: "18 is 6 times as many as 3." Some people call these **comparative statements**. See below for examples!

10 is *twice as much* as 5, so:

$$10 = 2 \times 5$$

8 is *2 times as big* as 4, so:

$$8 = 2 \times 4$$

20 is *5 times as much* as 4, so:

$$20 = 5 \times 4$$

12 is *3 times as many* as 4, so:

$$12 = 3 \times 4$$

100 is *10 times as big* as 10, so:

$$100 = 10 \times 10$$

35 is *7 times as many* as 5, so:

$$35 = 7 \times 5$$

Chapter 2

Pirates and Shiny Coins:
Skip Counting and the Multiplication Chart

Hidden Treasure . . . Skip Counting!

Do you like pirates? Hidden treasure? In this chapter, we'll learn a fun way to count treasure *and* understand multiplication better!

When you were younger, you might have learned to skip count, but we'll review it now, and I'll show you how it helps with understanding multiplication!

When we *skip count* by 2's, that means each number we say will be *2 more* than the one before it. (Say these out loud with me!)

Skip counting by 2's: 2, 4, 6, 8, 10 . . .

Now let's *skip count* by 3's—notice that each number we say will be *3 more* than the one before it:

Skip counting by 3's: 3, 6, 9, 12, 15 . . .

And we can do more!

Skip counting by 5's: 5, 10, 15, 20, 25 . . .

Skip counting by 10's: 10, 20, 30, 40, 50 . . .

Skip counting is counting by any whole number bigger than 1.

QUICK NOTE

When we **skip count** by a number, we're actually *adding* that number each time! For example, when skip counting by 4's, we're adding 4 each time: 4, 8, 12, 16 . . .

Let's see how skip counting looks on a number line—we actually "skip over" spaces as we count! Here, we're skip counting by 3's, and we do it 4 times: 3, 6, 9, 12. So we've shown $3 \times 4 = 12$. Notice that with each jump, we're counting 3 spaces between the tick marks (so we're counting the 12 <u>spaces</u>, not the tick marks themselves). And with each jump, we *add* 3 more, right? Yep, repeated addition like we saw on p. 16!

SKIP COUNTING BY 3'S   4 jumps!

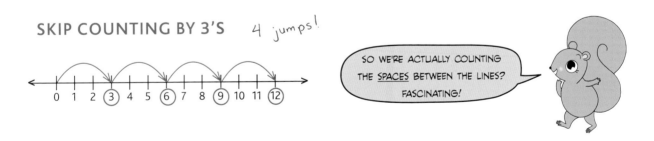

0 1 2 ③ 4 5 ⑥ 7 8 ⑨ 10 11 ⑫

SO WE'RE ACTUALLY COUNTING THE <u>SPACES</u> BETWEEN THE LINES? FASCINATING!

Skip counting and repeated addition are really the *same thing*, with two different names—and they are both great ways to find multiples!

Skip Counting with Coins to Find Multiples

Skip counting is a great tool for counting treasure . . . and learning multiplication facts!

Let's say we are pirates and we have 7 coins that are worth 5 cents each. We want to know how much it's worth in total. Since each coin is worth 5 cents, we could skip count by 5's for each of the 7 coins!

Let's count the treasure together! Touch each coin below and *skip count* out loud with me! "5, 10, 15, 20, 25, 30, **35**." Those are the first 7 *multiples* of 5, and this skip counting also tells us that $7 \times 5 = 35$, which is how much money we have: 35 cents!

5, 10, 15, 20, 25, 30, 35

We could also look at this as **repeated addition** on the number line:

SKIP COUNTING BY 5's! 7 jumps!

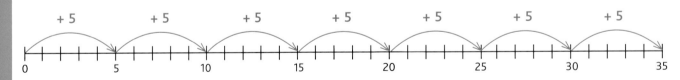

$$5 \times 7 = 35$$

Yep, we pirates have gotten ourselves 35 cents!

35 CENTS . . . ?

* The unit of money wasn't actually "cents"—it was, um, another name for a donkey that starts with "a." I'm not kidding. But if I told Mr. Mouse, we'd never hear the end of it. (No pun intended.)

Pirates and Shiny Coins: Skip Counting and the Multiplication Chart   33

The Multiplication Chart

Notice that the columns (and rows) of a multiplication chart allow us to easily skip count. For example, in the column with the "2" on top, we see 2, 4, 6 . . . etc.—they are all *multiples* of 2. Yep, we are adding 2 each time! And that's true for every column in the chart. Wanna skip count by 6's, see *repeated addition* in action, and find some multiples of 6? Then read the green "6" column: 6, 12, 18 . . . yep, we are adding 6 each time. Nice!

Multiplication Chart

×	1	2	3	4	5	6	7	8	9	10	11	12
1	1	2	3	4	5	6	7	8	9	10	11	12
2	2	4	6	8	10	12	14	16	18	20	22	24
3	3	6	9	12	15	18	21	24	27	30	33	36
4	4	8	12	16	20	24	28	32	36	40	44	48
5	5	10	15	20	25	30	35	40	45	50	55	60
6	6	12	18	24	30	36	42	48	54	60	66	72
7	7	14	21	28	35	42	49	56	63	70	77	84
8	8	16	24	32	40	48	56	64	72	80	88	96
9	9	18	27	36	45	54	63	72	81	90	99	108
10	10	20	30	40	50	60	70	80	90	100	110	120
11	11	22	33	44	55	66	77	88	99	110	121	132
12	12	24	36	48	60	72	84	96	108	120	132	144

And using this chart, we can quickly find *any* multiplication fact whose factors are less than or equal to 12. Of course, our goal in this book will be to *memorize* these multiplication facts, but first let's learn to use the chart. See the numbers in white, along the left and top border? Those will be the starting points—our two factors. For example, if we want to find the answer to 8 × 3 = ?, we could put one finger on the left 8, and another finger on the top 3. Then we slide our fingers in straight lines (staying inside the "8" row and the "3" column) until they run into each other at the answer: 24! And also notice that if we used the 8 on top and the 3 on the side, we'd still get 24—on a different part of the chart, of course! Try it now!

A *MAGICAL* MULTIPLICATION CHART. IT'S . . . SO . . . BEAUTIFUL . . . !

OH, MS. SQUIRREL. LET ME GET YOU A TISSUE. WAIT, THEY DIDN'T HAVE THOSE IN ANCIENT ROME.

Square Numbers on the Chart

On the previous page's multiplication charts, notice that the bold numbers on the diagonal are the products whose two factors are the same, like 3 × 3 = 9 or 5 × 5 = 25. These numbers are called "square numbers" or "perfect squares."

COOL! BUT WHY DO THEY CALL THEM SQUARE NUMBERS?

BECAUSE THE ARRAYS FOR THOSE MULTIPLICATION PROBLEMS *REALLY ARE* SQUARES. WHEN THE TWO FACTORS ARE THE SAME, IT MEANS THE NUMBERS OF ROWS AND COLUMNS ARE THE SAME, AND THAT MEANS THE ARRAY IS A <u>SQUARE</u> SHAPE-- JUST LIKE THE BASES OF THE PYRAMIDS IN EGYPT, WHICH, YEP, BECAME PART OF THE ROMAN EMPIRE!

TO BE CONTINUED...

The First Few Square Numbers

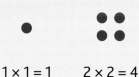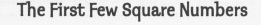

1 × 1 = 1   2 × 2 = 4   3 × 3 = 9   4 × 4 = 16   5 × 5 = 25

Another Way to Write Multiplication!

We can write multiplication problems either with the factors side by side (horizontally) or on top of each other (vertically). For example:

Written horizontally	Written vertically
6 × 3 = 18	$\begin{array}{r} 6 \\ \times\ 3 \\ \hline 18 \end{array}$
4 × 7 = 28	$\begin{array}{r} 4 \\ \times\ 7 \\ \hline 28 \end{array}$
10 × 10 = 100	$\begin{array}{r} 10 \\ \times\ 10 \\ \hline 100 \end{array}$

They're just two ways of saying the same thing!

Now let's practice using the multiplication chart!

GAME TIME!

Use the multiplication chart to complete the following multiplication facts.
I'll do the first one for you!

1.   7
    × 1
    ———
    ?

Let's Play: This answer might seem obvious, but let's use the chart anyway. We'll put one finger on the left white 7, and one on the top white 1. As we slide our fingers, we notice that we'd better not move the left finger too quickly or it will go too far! If that happens, it's okay—we can slide it back over again. The important part is that the left finger stays in its row and the top finger stays in its column. And we land on the 7!

Answer: 7 × 1 = 7

×	1	2	3	4	5	6	7	8	9	10	11	12
1	1	2	3	4	5	6	7	8	9	10	11	12
2	2	4	6	8	10	12	14	16	18	20	22	24
3	3	6	9	12	15	18	21	24	27	30	33	36
4	4	8	12	16	20	24	28	32	36	40	44	48
5	5	10	15	20	25	30	35	40	45	50	55	60
6	6	12	18	24	30	36	42	48	54	60	66	72
7	7	14	21	28	35	42	49	56	63	70	77	84
8	8	16	24	32	40	48	56	64	72	80	88	96
9	9	18	27	36	45	54	63	72	81	90	99	108
10	10	20	30	40	50	60	70	80	90	100	110	120
11	11	22	33	44	55	66	77	88	99	110	121	132
12	12	24	36	48	60	72	84	96	108	120	132	144

2.   5   3.   3
    × 4   × 7
    ———   ———
    ?   ?

4.   6   5.   5
    × 3   × 9
    ———   ———
    ?   ?

6.   4   7.   3   8.   9   9.   10
    × 8   × 5   × 9   × 2
    ———   ———   ———   ———
    ?   ?   ?   ?

10.   6   11.   9   12.   6   13.   8
    × 6   × 3   × 7   × 6
    ———   ———   ———   ———
    ?   ?   ?   ?

Keep going! ⟶

14.
$$\begin{array}{r} 7 \\ \times\ 7 \\ \hline ? \end{array}$$

15.
$$\begin{array}{r} 5 \\ \times\ 1 \\ \hline ? \end{array}$$

16.
$$\begin{array}{r} 8 \\ \times\ 7 \\ \hline ? \end{array}$$

17.
$$\begin{array}{r} 12 \\ \times\ 4 \\ \hline ? \end{array}$$

18.
$$\begin{array}{r} 7 \\ \times\ 4 \\ \hline ? \end{array}$$

19.
$$\begin{array}{r} 8 \\ \times\ 4 \\ \hline ? \end{array}$$

20.
$$\begin{array}{r} 8 \\ \times\ 8 \\ \hline ? \end{array}$$

21.
$$\begin{array}{r} 10 \\ \times\ 9 \\ \hline ? \end{array}$$

22.
$$\begin{array}{r} 4 \\ \times\ 6 \\ \hline ? \end{array}$$

23.
$$\begin{array}{r} 7 \\ \times\ 5 \\ \hline ? \end{array}$$

24.
$$\begin{array}{r} 11 \\ \times\ 6 \\ \hline ? \end{array}$$

25.
$$\begin{array}{r} 7 \\ \times\ 8 \\ \hline ? \end{array}$$

26.
$$\begin{array}{r} 9 \\ \times\ 7 \\ \hline ? \end{array}$$

27.
$$\begin{array}{r} 12 \\ \times\ 12 \\ \hline ? \end{array}$$

28.
$$\begin{array}{r} 11 \\ \times\ 8 \\ \hline ? \end{array}$$

29.
$$\begin{array}{r} 11 \\ \times\ 12 \\ \hline ? \end{array}$$

×	1	2	3	4	5	6	7	8	9	10	11	12
1	**1**	2	3	4	5	6	7	8	9	10	11	12
2	2	**4**	6	8	10	12	14	16	18	20	22	24
3	3	6	**9**	12	15	18	21	24	27	30	33	36
4	4	8	12	**16**	20	24	28	32	36	40	44	48
5	5	10	15	20	**25**	30	35	40	45	50	55	60
6	6	12	18	24	30	**36**	42	48	54	60	66	72
7	7	14	21	28	35	42	**49**	56	63	70	77	84
8	8	16	24	32	40	48	56	**64**	72	80	88	96
9	9	18	27	36	45	54	63	72	**81**	90	99	108
10	10	20	30	40	50	60	70	80	90	**100**	110	120
11	11	22	33	44	55	66	77	88	99	110	**121**	132
12	12	24	36	48	60	72	84	96	108	120	132	**144**

(Answers on page 220.)

Fun fact: In the movies, Queen Cleopatra of Egypt is shown to be incredibly beautiful. In real life, Cleopatra *wasn't* overly beautiful, but she was so smart and charming that people felt like they were under a spell—and some say that gave her lots of power as queen!

Chapter 3

Bagels and Hammers:
The Distributive Property

SOMETIMES THE BEST WAY TO TACKLE A BIG PROBLEM IS TO SPLIT IT INTO TWO SMALLER, EASIER PROBLEMS.

THAT'S NICE. BUT YOU KNOW WHAT I COULD GO FOR RIGHT ABOUT NOW? A BAGEL. MAYBE WITH SOME CREAM CHEESE ON IT?

OOH, AND HONEY!

ARE YOU GUYS EVEN PAYING ATTEN--?

I WONDER WHEN BAGELS WERE FIRST INVENTED. TIMES MACHINE, LET'S GO!

OKAY, FINE. HERE WE ARE IN 1683 AUSTRIA, WHERE MANY BELIEVE THE FIRST BAGELS WERE MADE TO HONOR THE KING OF POLAND FOR SAVING AUSTRIA FROM TURKISH INVADERS. THE KING LOVED TO RIDE HORSES, SO THE DOUGH WAS SHAPED LIKE A STIRRUP--WHICH IS *BÜGEL* IN GERMAN AND BECAME *BAGEL!*

STIRRUP

HEY, THIS IS REALLY HARD TO DO-- IT'S NOT WORKING!

UM, DUH--FIRST YOU NEED TO SPLIT THE BIG BAGEL IN TWO PARTS, THEN PUT THE HONEY ON BOTH. THIS IS MUCH EASIER BECAUSE THE TWO PARTS ARE FLAT. *THEN* YOU CAN PUT THE WHOLE THING BACK TOGETHER AGAIN.

SPLIT!

THANKS!

I KNOW A THING OR TWO ABOUT EATING.

HEY, GUESS WHAT? SPLITTING BAGELS IS ALSO GOING TO HELP US UNDERSTAND HOW THE *DISTRIBUTIVE PROPERTY* WORKS!

Splitting Bagels: The Distributive Property

The **Distributive Property** is a rule that allows us to rewrite multiplication problems as two <u>easier</u> problems—by splitting one of the factors! For example, we could solve 3 × 7 by first splitting the 7 into 2 & 5:

$$3 \times 7$$

Split! Split!

$$= 3 \times 2 + 3 \times 5$$

$$= 6 + 15 \quad \text{Then add 'em back together!}$$

$$= 21$$

And that's the answer: **3 × 7 = 21.**

The Distributive Property allows us to use <u>the multiplication facts we know</u> to more easily solve bigger multiplication problems!

"DECOMPOSING"? DOESN'T THAT MEAN "ROTTING"? BELIEVE ME, I'D EAT THE BAGEL WAAAAY BEFORE IT GOT CLOSE TO ROTTING.

Splitting = Decomposing

Instead of "splitting" the factors into two smaller numbers, some textbooks call this "decomposing" the factors. It's just two ways of saying the same thing!

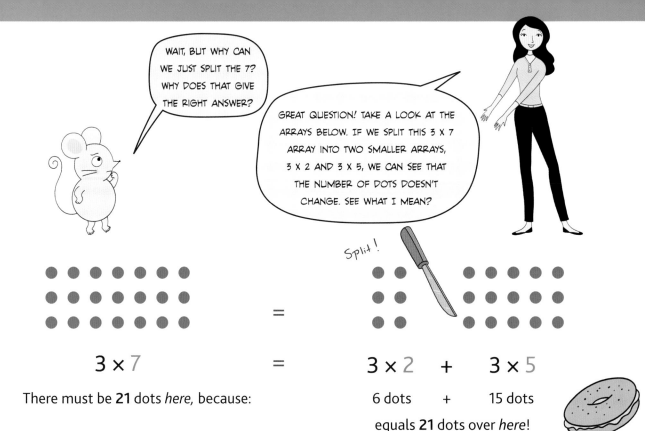

$$3 \times 7 = 3 \times 2 + 3 \times 5$$

There must be **21** dots *here,* because:

6 dots + 15 dots

equals **21** dots over *here!*

See? We didn't *change* the number of dots—we just split them up! Since 3 × 2 = **6** and 3 × 5 = **15**, we can add those together (6 + 15 = **21**), and we've learned that the total number of dots for 3 × 7 must be **21**, too.

Peas and Parentheses

Parentheses are a great little tool (aren't they cute?) to separate out words or numbers. In the above example, we could have written: 3 × 7 = (3 × 2) + (3 × 5). Parentheses can make things look a little more complicated, but don't let them scare you! They help tell us what part of the math problem to solve first (see Chapter 6 for more!), and they do a great job of keeping things separate, like when you don't want your peas to touch your mashed potatoes.

The Distributive Property for Breaking Off a Ten—with a Hammer!

Multiplication by 10 is super easy—<u>we just add a zero!</u> For example, 5 × 10 = **50**, and 9 × 10 = **90**—see what I mean? (Adding a zero moves the *place value* by one spot—see more on p. 152.) So sometimes when we break up a number into "smaller, easier" numbers, we might choose 10 as one of them. It's not very small, but it's REALLY easy to multiply with! For example, let's try 4 × 12 = ? by breaking 12 into 2 and 10:

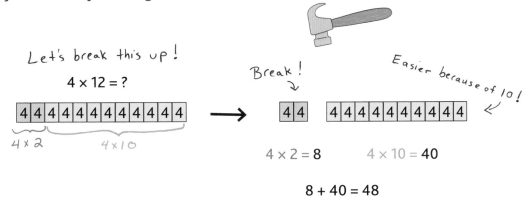

So 4 × 12 = **48**!

Did you notice how 4 × 12 is the same as 4 × 2 + 4 × 10? In other words, 4 × 12 = 4 × 2 + 4 × 10. And we can even use parentheses, just to keep everything separate: 4 × 12 = (4 × 2) + (4 × 10). Nobody wants their peas touching their mashed potatoes, after all!

Multiplication Made Easier— with the Distributive Property!

STEP BY STEP:

The "Split the Bagel" or "Hammer" Trick

Step 1. Split (decompose!) the big factor into two smaller, friendlier numbers, and write our two (new, easier!) multiplication problems, using parentheses.

Step 2. Do both (new, easier!) multiplication problems.

Step 3. Add *those* answers together for the final answer. Done!

Let's say we want to multiply 3 × 12, but can't remember that fact.

The Distributive Property tells us we can find the answer like this:

Let's use the **Step by Step** above!	3 × 12
Step 1. Split the 12 into 2 & 10 and rewrite it as two (easier!) problems with parentheses.	(3 × 2) + (3 × 10)
Step 2. Solve the two (easier!) problems.	(3 × 2) = 6 and (3 × 10) = 30
Step 3. Add the two answers back together for the final answer!	6 + 30 = **36**

Not so bad, right? Let's practice some of this splitting! You can think of splitting bagels or breaking numbers with a hammer—whichever you prefer!

GAME TIME!

These multiplication problems were broken up into two easier ones. Figure out what the missing number is, do the two easier problems by using the multiplication chart below (see p. 34 for a refresher), and add 'em together for the final answer. I'll do the first one for you!

1. 9 × 13 = (9 × 10) + (9 × ___?___)

Let's Play: Hmm, if we took a hammer and broke the 13, and we know one piece is 10, that means the other piece must be *3*, since 10 + 3 = 13, right? Great! So we have (9 × 10) + (9 × 3). The multiplication chart below tells us that 9 × 10 = 90, and 9 × 3 = 27, so now we just add 'em up: 90 + 27 = 117. Great!

Answer: 9 × 13 = (9 × 10) + (9 × 3) = 117

2. 7 × 13 = (7 × 3) + (7 × ___?___)

3. 5 × 15 = (5 × ___?___) + (5 × 10)

4. 6 × 14 = (6 × 10) + (6 × ___?___)

5. 4 × 16 = (4 × 10) + (4 × ___?___)

6. 4 × 17 = (4 × 7) + (4 × ___?___)

7. 8 × 12 = (8 × 2) + (8 × ___?___)

×	1	2	3	4	5	6	7	8	9	10	11	12
1	1	2	3	4	5	6	7	8	9	10	11	12
2	2	4	6	8	10	12	14	16	18	20	22	24
3	3	6	9	12	15	18	21	24	27	30	33	36
4	4	8	12	16	20	24	28	32	36	40	44	48
5	5	10	15	20	25	30	35	40	45	50	55	60
6	6	12	18	24	30	36	42	48	54	60	66	72
7	7	14	21	28	35	42	49	56	63	70	77	84
8	8	16	24	32	40	48	56	64	72	80	88	96
9	9	18	27	36	45	54	63	72	81	90	99	108
10	10	20	30	40	50	60	70	80	90	100	110	120
11	11	22	33	44	55	66	77	88	99	110	121	132
12	12	24	36	48	60	72	84	96	108	120	132	144

(Answers on page 220.)

Chapter 4

Bananas for Monkeys and Dinosaur Bones:
Intro to Division

Division: Fair Sharing . . . Bananas!

Division is when we split a number into *equal* parts or groups. For example, 12 divided by 3 is **4**, because if we split 12 into 3 equal groups, there will be **4** in each group.

12 divided by 3 equals 4

$12 \div 3 = 4$

Fair Sharing vs. Measurement

As we've seen, division can be thought of as "fair sharing":

HEY, MR. MOUSE, IF WE HAVE 12 BANANAS AND 4 MONKEYS, HOW MANY BANANAS DO THEY EACH GET IF THEY'RE *SHARING FAIRLY?*

WELL, SINCE 12 ÷ 4 = 3, THE MONKEYS COULD GET 3 BANANAS EACH. BUT I THINK I SHOULD GET SOME.

And division can also be thought of as "measurement":

HEY, MS. SQUIRREL, IF WE HAVE 12 BANANAS AND EACH MONKEY WILL GET 3 BANANAS, *HOW MANY MONKEYS CAN GET BANANAS?*

SINCE 12 ÷ 3 = 4, WE COULD GIVE A 3-BUNCH OF BANANAS TO 4 MONKEYS! BUT I THINK ALL MONKEYS EVERYWHERE SHOULD GET BANANAS, NOT JUST 4 OF THEM.

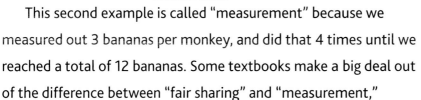

This second example is called "measurement" because we measured out 3 bananas per monkey, and did that 4 times until we reached a total of 12 bananas. Some textbooks make a big deal out of the difference between "fair sharing" and "measurement,"

but when it comes down to it, they're both just about splitting things up into equal groups, after all—division! And a great way to "see" division is with *arrays*.

Return of the Arrays

Remember the array of ants on p. 15? We showed a total of 20 ants, as 4 groups of 5 each, for the multiplication problem $4 \times 5 = 20$. But this also shows the *division* problem **20 ÷ 4 = 5**. After all, to divide 20 into 4 groups, we could circle 4 equal groups—and that gives us 5 in each group: "20 divided by 4 equals 5"!

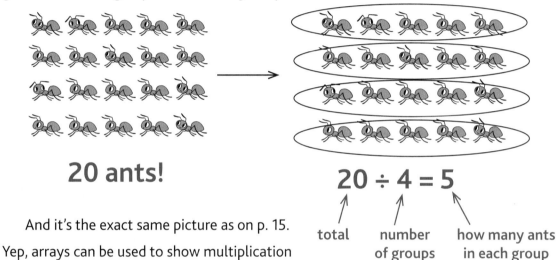

20 ants!

And it's the exact same picture as on p. 15. Yep, arrays can be used to show multiplication *and* division problems.

$$20 \div 4 = 5$$

total   number of groups   how many ants in each group

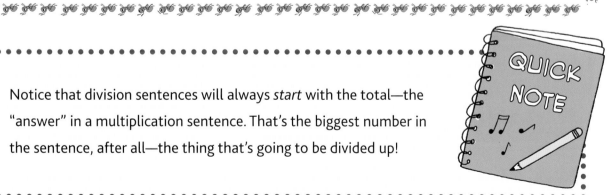

Notice that division sentences will always *start* with the total—the "answer" in a multiplication sentence. That's the biggest number in the sentence, after all—the thing that's going to be divided up!

Let's look at some arrays and figure out the division problems ourselves!

GAME TIME!

Use the pictures to complete the division sentences.
I'll do the first one for you!

1.
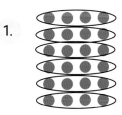

24 ÷ 6 = __?__

Let's Play: It looks like we've got 24 dots total, and we are dividing them into 6 equal groups. How many dots are in each group? We see 4 in each group!

Answer: 24 ÷ 6 = 4

2.

___?___ ÷ 4 = 3

3.

15 ÷ __?__ = 5

4.

___?___ ÷ 3 = __?__

5.

10 ÷ 5 = __?__

6.

16 ÷ __?__ = 4

7.

16 ÷ __?__ = __?__

8.

___?___ ÷ 4 = __?__

9.

___?___ ÷ __?__ = 7

10.

___?___ ÷ __?___ = __?___

(Answers on page 220.)

Remember on p. 16 when we saw that multiplication is just *repeated addition,* like how 6 × 4 is the same as 6 + 6 + 6 + 6? Let's see how that looks on a number line.

$$6 \times 4 = \underline{\ ?\ } \qquad 6 + 6 + 6 + 6 = 24 \qquad \text{So, } 6 \times 4 = 24$$

We add 6,
4 times

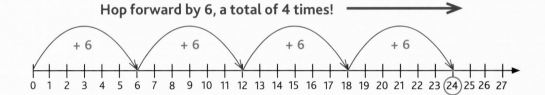

In that same way, we can think of division as *repeated subtraction.* For example, if we wanted to solve 24 ÷ 6, we could start with 24 and begin subtracting 6. *How many times* do we subtract 6 until we get to zero? Well, that number will be our answer!

$$24 \div 6 = \underline{\ ?\ } \qquad 24 - 6 - 6 - 6 - 6 = 0 \qquad \text{So, } 24 \div 6 = 4$$

We subtract 6,
4 times

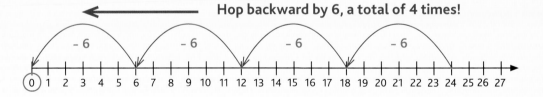

Just like 4 hops forward gets us to 24, we can <u>undo that</u> by hopping *backward* 4 times to get back to zero. So, division just *undid* the multiplication on the number line, see? Speaking of "undoing" . . .

Untying Our Shoes!
Multiplication and Division as Inverse Operations

In *Do Not Open This Math Book,* we learned all about addition and subtraction. Now that you're a little older, I want you to notice something cool about these two operations: They *undo* each other! Let's say we start with 9, and we <u>add 1</u> to it, and get 10. Then we <u>subtract 1</u>, and we get right back to 9.

When we subtracted 1, we *undid* the 1 that we added, right? So we say that subtraction can *undo* addition. And addition can *undo* subtraction, too. This is because addition and subtraction are **inverse operations.**

Inverse operations are two operations that "undo" each other.

For example, *addition and subtraction* are inverse operations, because if we start with 9 and <u>subtract 4</u>, we get 5—but we can *undo* that by <u>adding 4</u> and get right back to the 9.

Multiplication and division are also inverse operations. For example, if we start with 6 and <u>divide by 3</u>, we get 2—but we can *undo* that by <u>multiplying by 3</u> and get right back to the 6.

"Inverse" means something very specific in math—it's when one thing UNDOES another, like *un*tying your shoes, taking *off* your hat, or *deleting* letters you've just typed. And "reverse" is more like, well, "backward." Check it out:

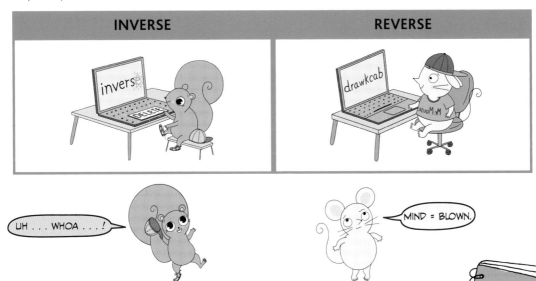

There are more ways division can be written, too. See p. 60 for how we write division using "dividing houses," like:

$$3 \overline{)\,12\,}^{4}$$

And we can also use a fraction: $\dfrac{12}{3} = 4$

We won't be using fractions in this book, but when you're ready, check out p. 42 in my book *Math Doesn't Suck* to learn how fractions are actually *division problems*— in disguise!

Fact Families . . . and Dinosaurs!

Because multiplication and division are inverse operations, there are **fact families** for them—just like with addition and subtraction! (See p. 31 in *Do Not Open This Math Book*.)

A **fact family** for multiplication and division is a group of multiplication and division facts that all use the same three numbers! Here are some examples of fact families:

2 × 8 = 16	3 × 4 = 12	7 × 10 = 70
8 × 2 = 16	4 × 3 = 12	10 × 7 = 70
16 ÷ 2 = 8	12 ÷ 3 = 4	70 ÷ 7 = 10
16 ÷ 8 = 2	12 ÷ 4 = 3	70 ÷ 10 = 7

If the two factors in the multiplication fact are the same (so the product will be a *square number*), then there will only be *two* facts in the fact family, not the usual four. Here are some examples:

3 × 3 = 9	8 × 8 = 64
9 ÷ 3 = 3	64 ÷ 8 = 8

Fact families are great, because this means there's *less* to memorize. For example, if we know that 2 × 3 = 6, then we automatically know the other multiplication fact in the "family" by just switching the order of the factors: 3 × 2 = 6. We also know the division facts in this fact family—we just have to make sure to put the biggest number first (because it's the thing that will get divided up), so we know 6 ÷ 3 = 2 and 6 ÷ 2 = 3. See? There's less to memorize!

 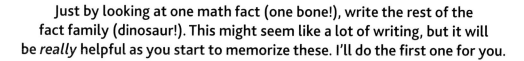

GAME TIME!

Just by looking at one math fact (one bone!), write the rest of the fact family (dinosaur!). This might seem like a lot of writing, but it will be *really* helpful as you start to memorize these. I'll do the first one for you.

1. $8 \times 9 = 72$

Let's Play: That's a big bone! Okay, most fact families have two multiplication facts and two division facts, right? To get the other multiplication fact, we can just reverse the order and get $9 \times 8 = 72$. How do we make the division facts? Hmm, well, we know the 72 has to come first, because it's the big total and so that's what we'll be dividing up. First let's divide 72 into 8 equals parts; that would be $72 \div 8 = 9$. Then we can divide 72 into 9 equal parts: $72 \div 9 = 8$. We got all four facts—yippee!

Answer:

$8 \times 9 = 72$ $72 \div 8 = 9$

$9 \times 8 = 72$ $72 \div 9 = 8$

2. $2 \times 3 = 6$
3. $10 \times 9 = 90$
4. $7 \times 8 = 56$
5. $6 \times 5 = 30$

6. $9 \times 2 = 18$
7. $6 \times 7 = 42$
8. $4 \times 1 = 4$
9. $8 \times 6 = 48$

10. $5 \times 11 = 55$
11. $8 \times 4 = 32$
12. $6 \times 6 = 36$
    Hint: This is a square number!
13. $63 \div 7 = 9$

14. $28 \div 7 = 4$
15. $84 \div 7 = 12$
16. $54 \div 9 = 6$
17. $18 \div 6 = 3$

18. $132 \div 11 = 12$
19. $27 \div 3 = 9$
20. $100 \div 10 = 10$
21. $12 \times 12 = 144$

(Answers on page 220.)

How Many Facts in a Fact Family?

By the way, some textbooks like to write eight facts for fact families, instead of the typical four that we've been doing. For example, some textbooks would count $16 = 2 \times 8$ as well as $2 \times 8 = 16$, so the entire thing would look like this:

$$2 \times 8 = 16 \qquad 16 = 2 \times 8$$
$$8 \times 2 = 16 \qquad 16 = 8 \times 2$$
$$16 \div 2 = 8 \qquad 8 = 16 \div 2$$
$$16 \div 8 = 2 \qquad 2 = 16 \div 8$$

And for a square fact family, we'll use just two facts, like $3 \times 3 = 9$ and $9 \div 3 = 3$. But some textbooks would use four facts, like this:

$$3 \times 3 = 9 \qquad 9 = 3 \times 3$$
$$9 \div 3 = 3 \qquad 3 = 9 \div 3$$

It seems like a lot of extra writing to me, but do what your teacher says!

What Now? Rethinking Division Problems as Multiplication!

Let's face it: Multiplication problems are easier than division problems. The great news is that we can rethink any division fact as a multiplication fact, like going upside down to see a room differently!

For example, say we know **4 × 3 = 12**. If someone comes along and asks us to answer the division problem **12 ÷ 4 = ?**, we could "stand on our heads" and rethink this as a *multiplication* fact (with something missing!) and say, "**4 times *what* equals 12?**" In other words, we can think: "**4 × ? = 12**." That's a multiplication fact with something missing, right? And the thing that goes where the "?" is will be our division answer.

Looking at 4 × ? = 12, we'd be like, "Oh! I know that one! It's just 4 × <u>3</u> = 12." And there's our answer: **12 ÷ 4 = <u>3</u>**. Ta-da!

THE SAME ROOM LOOKS PRETTY DIFFERENT UPSIDE DOWN, DOESN'T IT?

I SEE WHAT YOU MEAN.

Let's practice!

GAME TIME!

Answer the following division problems by first writing them as multiplication facts with something missing, and then using the bank of multiplication facts to help. Remember that the big number that's getting divided up will be the product in the multiplication fact. I'll do the first one for you!

Multiplication facts to use:

$4 \times 6 = 24$ $7 \times 7 = 49$ $5 \times 4 = 20$ $6 \times 2 = 12$

$7 \times 8 = 56$ $9 \times 7 = 63$ $11 \times 10 = 110$

1. $63 \div 9 = \underline{\ ?\ }$

Let's Play: The first thing we should do is "stand on our heads" and write this division problem as a <u>multiplication</u> fact with *something missing*. The big number, 63, must be the <u>product</u> in the multiplication fact, right? So we could think, "9 times *what* equals 63?" or 9 × ? = 63. Then we can look at our bank of multiplication facts and find 9 × 7 = 63, so the "what" is 7. Great!

Answer: $9 \times \underline{\ ?\ } = 63$ and $63 \div 9 = 7$

2. $49 \div 7 = \underline{\ ?\ }$

Hint: 7 × $\underline{\ ?\ }$ = 49

3. $12 \div 2 = \underline{\ ?\ }$

Hint: 2 × $\underline{\ ?\ }$ = 12

4. $12 \div 6 = \underline{\ ?\ }$

(No more hints!)

5. $20 \div 5 = \underline{\ ?\ }$

6. $20 \div 4 = \underline{\ ?\ }$

7. $24 \div 4 = \underline{\ ?\ }$

8. $24 \div 6 = \underline{\ ?\ }$

9. $63 \div 7 = \underline{\ ?\ }$

10. $110 \div 10 = \underline{\ ?\ }$

11. $110 \div 11 = \underline{\ ?\ }$

12. $56 \div 8 = \underline{\ ?\ }$

13. $56 \div 7 = \underline{\ ?\ }$

(Answers on page 220.)

Prehistoric Cave Houses: Another Way to Write Division

The Parts of Division

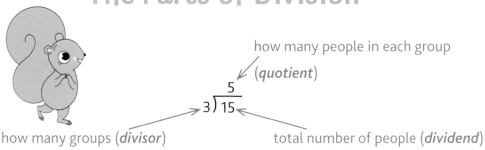

how many people in each group (*quotient*)

$$3\overline{)15}^{\,5}$$

how many groups (*divisor*)

total number of people (*dividend*)

And we can read the above problem like this: "Fifteen divided by three is five."

The Parts of Division—Divisor, Dividend, Quotient

In case you have to memorize the names *divisor*, *dividend,* and *quotient,* here's a fun way to think about it. Instead of a cave house, imagine it's a real house, and there's a dog outside, looking at a steak on the table. He's got his "eye" on the steak—he's the **div-EYE-sor** (that's how you pronounce *divisor*). His "end" goal is the steak—the **div-id-END**—and the little girl on top of the house is saying "Sparky! You'll get your dinner soon!" She's being quoted—she's the **Quotient**.

See? Divisor, dividend, quotient!

Watch the "Math Doesn't Suck: How to Focus" episode at **mckellarmath.com/fun-links** to see this as a video!

These are all ways of saying the same thing:

$24 \div 3 = 8$ "24 divided by 3 equals 8."

$$3\overline{)24}\,^{\,8}$$ "3 goes into 24, 8 times."

"The number of *groups of 3* found in 24 is 8."

Tennis, Anyone? Dividing "into" Numbers

When looking at a "dividing house" problem like this (which some people call *long division*), I like to think, "How many *groups of 3* fit inside 15?" or "How many times does 3 *go into* 15?" It seems to match the picture, like that 3 is just waiting to go inside the cave to see how many of them fit.

$$3\overline{)15}$$

Let's pretend the "3" is a stack of 3 tennis balls, and the "15" is a set of tennis ball containers that fit 3 balls each. Then we're asking, "How many times do *3 balls* fit?"

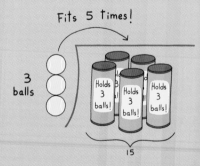

And of course, the answer is 5! This is one fun way to think about what it means to divide a small number "into" another number. There are lots of ways to think about division—use whatever you like the most!

Using the Multiplication Chart to Find Division Answers

Remember how we used the multiplication chart on p. 34 to find multiplication answers? Well, that same chart can be used to find division answers, too!

×	1	2	3	4	5	6	7	8	9	10	11	12
1	1	2	3	4	5	6	7	8	9	10	11	12
2	2	4	6	8	10	12	14	16	18	20	22	24
3	3	6	9	12	15	18	21	24	27	30	33	36
4	4	8	12	16	20	24	28	32	36	40	44	48
5	5	10	15	20	25	30	35	40	45	50	55	60
6	6	12	18	24	30	36	42	48	54	60	66	72
7	7	14	21	28	35	42	49	56	63	70	77	84
8	8	16	24	32	40	48	56	64	72	80	88	96
9	9	18	27	36	45	54	63	72	81	90	99	108
10	10	20	30	40	50	60	70	80	90	100	110	120
11	11	22	33	44	55	66	77	88	99	110	121	132
12	12	24	36	48	60	72	84	96	108	120	132	144

For example, let's say we wanted to solve $56 \div 8 = ?$. We could first find the 8 on the left. See it? Now put your finger on it. Next, let's move to the right, staying in our row or column, until we see the 56. Now we go straight up from that 56, and we'll see 7. And that tells us that $8 \times 7 = 56$—in other words, $56 \div 8 = 7$. Ta-da! By the way, we also could have started with the top 8, moved down to the other 56, and then looked left to see the 7. Either way works!

Let's practice using our multiplication chart and our little dividing "cave" houses!

Let's practice! Rewrite each of these problems with dividing houses, solve them using the multiplication chart, and then practice saying the division fact out loud. Remember, the biggest number goes inside the cave. I'll do the first one for you!

1. $84 \div 12 =$ _?_

Let's Play: Okay, first we're supposed to write this with a dividing house. Hmm, the biggest number goes inside the cave, so we'll put the 84 inside, and the 12 goes on the outside, like this: $12\overline{)84}$. Next, we need to solve it. If we find the 84 on the chart, we can move our fingers up and to left to find 12 and **7**, so that means 7 is our answer! Great! So we'll write:

$$12\overline{)84}^{\,7}$$

And finally, let's look at our little cave problem and say it out loud: "Eighty-four divided by twelve equals seven." Done!

Answer: $12\overline{)84}^{\,7}$ (and saying it out loud!)

2. $9 \div 3 =$ _?_   3. $14 \div 7 =$ _?_   4. $70 \div 7 =$ _?_   5. $100 \div 10 =$ _?_

6. $64 \div 8 =$ _?_   7. $28 \div 4 =$ _?_   8. $32 \div 8 =$ _?_   9. $121 \div 11 =$ _?_

10. $63 \div 9 =$ _?_   11. $48 \div 6 =$ _?_   12. $45 \div 5 =$ _?_   13. $42 \div 7 =$ _?_

For these problems, just solve them and say them out loud!
(No need to rewrite them, since they are already in "cave" form.)

14. $9\overline{)54}$?

15. $3\overline{)15}$?

16. $10\overline{)120}$?

17. $9\overline{)81}$?

18. $6\overline{)36}$?

19. $2\overline{)18}$?

20. $5\overline{)25}$?

21. $12\overline{)144}$?

22. $11\overline{)121}$?

23. $7\overline{)56}$?

24. $7\overline{)63}$?

25. $12\overline{)132}$?

×	1	2	3	4	5	6	7	8	9	10	11	12
1	**1**	2	3	4	5	6	7	8	9	10	11	12
2	2	**4**	6	8	10	12	14	16	18	20	22	24
3	3	6	**9**	12	15	18	21	24	27	30	33	36
4	4	8	12	**16**	20	24	28	32	36	40	44	48
5	5	10	15	20	**25**	30	35	40	45	50	55	60
6	6	12	18	24	30	**36**	42	48	54	60	66	72
7	7	14	21	28	35	42	**49**	56	63	70	77	84
8	8	16	24	32	40	48	56	**64**	72	80	88	96
9	9	18	27	36	45	54	63	72	**81**	90	99	108
10	10	20	30	40	50	60	70	80	90	**100**	110	120
11	11	22	33	44	55	66	77	88	99	110	**121**	132
12	12	24	36	48	60	72	84	96	108	120	132	**144**

(Answers on page 220.)

Chapter 5

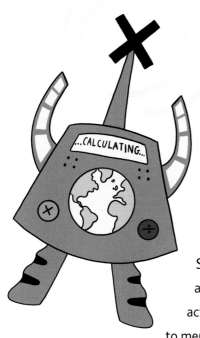

The Core of the Times Machine:
Tricks and Stories for
Memorizing the Multiplication Facts!

So now that we've got a good feeling for what multiplication and division are, it's time to go deep into the Core of the Times Machine and learn the actual multiplication and division facts! This "machine" chapter will help you to memorize your facts once and for all! Don't worry if it takes a while to learn them all—this stuff takes "times.". . .

I SEE WHAT YOU DID THERE.

A Quick List of Multiplication Tricks!

In the following pages, I'll show you easy ways to remember *all* the facts—but first, here are some of my favorite tricks, all together!

To Multiply By:	Here's a Trick:	For Example:
2	Double the number! Remember our doubles from addition—it's the same thing!	$7 \times 2 = 14$ because $7 + 7 = 14$
3	Double the number, and then *add one more* of that same number!	$8 \times 3 = 24$ because $8 + 8 = 16$ (double the 8) $16 + 8 = 24$ (add one more 8!)

4	Double the number, and then double it again!	$7 \times 4 = 28$ because $7 + 7 = 14$ and $14 + 14 = \mathbf{28}$
5	Multiples of 5 *always* end in 0 or 5, and it's really easy to skip count to get these answers! ——————— *OR* We can multiply by 10, and then divide by 2!	$4 \times 5 = 20$ We can skip count by 5's, *four* times: 5, 10, 15, **20** ——————— *OR* We could first multiply by 10: $4 \times 10 = 40$ and then divide by 2: $40 \div 2 = \mathbf{20}$
9	Use the 9's Fingers Trick (explained in detail on p. 110) ——————— *OR* The "What if the 9 were a 10?" trick: We can multiply by 10, but then, oops, that's too much! So we then *subtract* the number we were multiplying.	6 3 Bend down 7th finger for $9 \times 7 = \mathbf{63}$ ——————— *OR* $10 \times 7 = 70$, but oops, that's too many 7's! So we subtract one 7: $70 - 7 = \mathbf{63}$.
10	To multiply 10 times any whole number, we need to change the *place value* by one spot, so we can just *stick a zero on the end*!	$8 \times 10 = 80$ When we stick a 0 on the 8, we get 80!
11	If we're multiplying 11 times a single-digit number, we just *repeat* that number— once in the tens place and once in the ones place.	$11 \times 8 = 88$ ↗ ↖ tens ones place place
12	Since $12 = 10 + 2$: We multiply the original number by 10, and also multiply the original number by 2. Then we add the two answers together!	$12 \times 7 = 84$ We just do this: $10 \times 7 = 70$ $2 \times 7 = 14$ and $70 + 14 = \mathbf{84}$

These tricks are meant to be helpful—use the ones you like, don't worry about the ones you don't like, and take your time! Learning the multiplication and division facts can take a while, so don't rush through these. For example, maybe take a week or two with the 3's section and all the cute little stories in them. Then test yourself with the Game Time sections here and also try the resources at **TheTimesMachine.com** (like multiplication facts you can tape to your bathroom mirror!), and get good at the 3's before moving on to the 4's section. If you try to do all of it at once, it could be frustrating, which nobody wants. Let's make this fun instead!

Why Should I Learn to Drive a Car?

Learning the Multiplication Facts

We've learned that big multiplication facts can be figured out by breaking them into smaller, easier problems like we did on p. 41, so why should we bother memorizing ALL the 0–12 facts? That's like saying, "I know how to walk, so why should I learn how to drive?" Knowing the multiplication facts will make you SO much faster at your math homework (and tests!) in the next few years, and not just for actual multiplication and division. For example, once you start simplifying fractions, you'll *really* need to know those facts (like we'll do in my book *Math Doesn't Suck!*). Memorizing your multiplication facts will make you feel like a superstar—you'll be like, oh yeah, I totally know $6 \times 8 = 48$, bam!

THE ZEROS

It's nice that we get to start out with the easiest multiplication facts: the zeros! ANYTHING times zero is zero. Always. It's like zero sucks all numbers into its zero-ness and turns them into zero. So $5 \times 0 = 0$, and $46 \times 0 = 0$, you name it.

The "0" Multiplication Facts

$0 \times 1 = 0$	$0 \times 2 = 0$	$0 \times 3 = 0$	$0 \times 4 = 0$	$0 \times 5 = 0$	$0 \times 6 = 0$
$1 \times 0 = 0$	$2 \times 0 = 0$	$3 \times 0 = 0$	$4 \times 0 = 0$	$5 \times 0 = 0$	$6 \times 0 = 0$
$0 \times 7 = 0$	$0 \times 8 = 0$	$0 \times 9 = 0$	$0 \times 10 = 0$	$0 \times 11 = 0$	$0 \times 12 = 0$
$7 \times 0 = 0$	$8 \times 0 = 0$	$9 \times 0 = 0$	$10 \times 0 = 0$	$11 \times 0 = 0$	$12 \times 0 = 0$

And by the way, $0 \times 0 = 0$, too. Not that this comes as a surprise.

 ## Why We Can't Divide by Zero: The Case of the Missing Monkey!

Remember on p. 47, when we asked: "If we have 8 bananas and there are 4 monkeys to share them equally, how many bananas will each monkey get?" That's the division sentence 8 ÷ 4 = 2. In other words, 8 bananas divided among 4 monkeys means 2 bananas for each monkey. What if instead we asked, "If we have 8 bananas and 0 monkeys to share them equally, *how many bananas will each monkey get?*" That would be the division sentence: 8 ÷ 0 = ? But—what? There are no *monkeys* to give bananas to! We can't divide up 8 bananas among 0 monkeys. It simply doesn't *mean* anything. See? We should never try to divide by zero. It's just bananas!

Dividing Zero by Other Numbers

Although we can't divide by 0, we *can* divide 0 *by other numbers* . . . and the answer is always zero. For example, we can ask the question: "If we have no bananas, and 4 monkeys share them equally, how many bananas will each monkey get?" In other words: 0 ÷ 4 = ? If we have no bananas, then the answer of course is that the monkeys get 0 bananas each: **0 ÷ 4 = 0**. It's a little sad that they don't get any bananas, but at least it makes *sense*. See what I mean? . . .

Okay, I'll go get them some bananas now.

The "0" Division Facts

0 ÷ 1 = 0	0 ÷ 2 = 0	0 ÷ 3 = 0	0 ÷ 4 = 0	0 ÷ 5 = 0	0 ÷ 6 = 0
0 ÷ 7 = 0	0 ÷ 8 = 0	0 ÷ 9 = 0	0 ÷ 10 = 0	0 ÷ 11 = 0	0 ÷ 12 = 0

Just to recap, division facts with zero like these are okay:

 $0 \div 7 = 0$ $7 \overline{)\,0\,}^{\,0}$

And these are NOT okay—they are not allowed, and don't even *mean* anything.

 $7 \div 0 = ???$ $0 \overline{)\,7\,}^{\,???}$

For thousands of years, people did math *without zero*. In AD 350, in what is now Mexico, the Mayans invented a version of zero for their calendar systems. Then around AD 628, an Indian mathematician named Brahmagupta developed a symbol for zero and used it in *equations* for the first time. But Europe said, "Thanks for nothing, Brah!" Some people thought "zero" was somehow ungodly, so it wasn't widely accepted in Europe until the 1600s.

THE ONES

In some ways, the ones are even easier than the zeros. . . . You see, 1 times any number is just that *same number*. So $1 \times 3 = 3$, and $70 \times 1 = 70$, etc.

I mean, for 1×5, we only have 1 row and 5 columns, right? So the "array" is just going to be a line of 5 dots! Or for 5×1, we have 5 rows but only 1 column, so our "array" is just an up-and-down line of 5 dots! See what I mean?

$1 \times 5 = 5$

$5 \times 1 = 5$

And that's all there is to it!

The "1" Multiplication Facts

1 × 1 = 1 1 × 1 = 1	1 × 2 = 2 2 × 1 = 2	1 × 3 = 3 3 × 1 = 3	1 × 4 = 4 4 × 1 = 4	1 × 5 = 5 5 × 1 = 5	1 × 6 = 6 6 × 1 = 6
1 × 7 = 7 7 × 1 = 7	1 × 8 = 8 8 × 1 = 8	1 × 9 = 9 9 × 1 = 9	1 × 10 = 10 10 × 1 = 10	1 × 11 = 11 11 × 1 = 11	1 × 12 = 12 12 × 1 = 12

And division by 1 is really easy, too: If we divide any number by 1, we get that same number again! So 5 ÷ 1 = 5, and 87 ÷ 1 = 87, etc. It makes sense if you think about it: Let's say we have 8 bananas and 1 monkey shares the bananas equally—yep, that 1 monkey gets all 8 bananas! That's 8 ÷ 1 = 8.

Let's talk about the reverse of that, too: If we divide any number by itself, we get 1 (except zero—see p. 70!). Think about it: If we have 8 bananas and 8 monkeys to share them equally, they'd each get 1 banana, right? That's 8 ÷ 8 = 1. Ta-da!

The "1" Division Facts

1 ÷ 1 = 1 1 ÷ 1 = 1	2 ÷ 1 = 2 2 ÷ 2 = 1	3 ÷ 1 = 3 3 ÷ 3 = 1	4 ÷ 1 = 4 4 ÷ 4 = 1	5 ÷ 1 = 5 5 ÷ 5 = 1	6 ÷ 1 = 6 6 ÷ 6 = 1
7 ÷ 1 = 7 7 ÷ 7 = 1	8 ÷ 1 = 8 8 ÷ 8 = 1	9 ÷ 1 = 9 9 ÷ 9 = 1	10 ÷ 1 = 10 10 ÷ 10 = 1	11 ÷ 1 = 11 11 ÷ 11 = 1	12 ÷ 1 = 12 12 ÷ 12 = 1

× # THE TWOS ×

Multiplying by 2 is also called *doubling*.

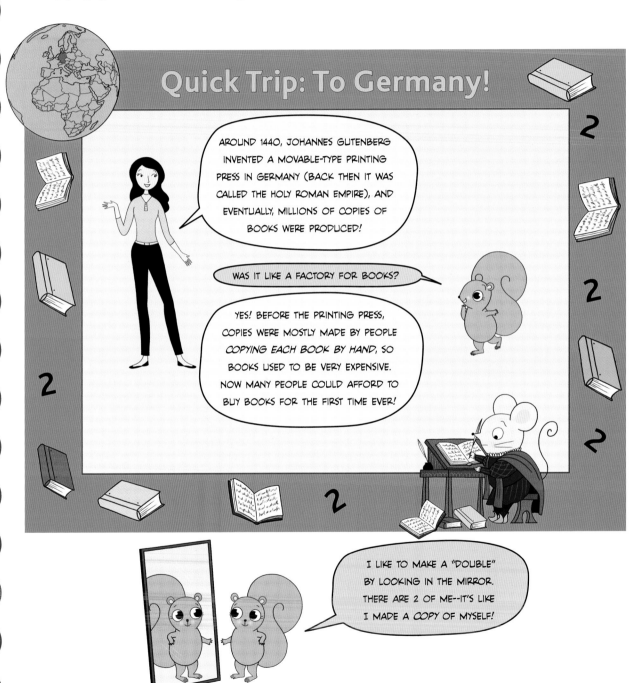

Quick Trip: To Germany!

AROUND 1440, JOHANNES GUTENBERG INVENTED A MOVABLE-TYPE PRINTING PRESS IN GERMANY (BACK THEN IT WAS CALLED THE HOLY ROMAN EMPIRE), AND EVENTUALLY, MILLIONS OF COPIES OF BOOKS WERE PRODUCED!

WAS IT LIKE A FACTORY FOR BOOKS?

YES! BEFORE THE PRINTING PRESS, COPIES WERE MOSTLY MADE BY PEOPLE *COPYING EACH BOOK BY HAND*, SO BOOKS USED TO BE VERY EXPENSIVE. NOW MANY PEOPLE COULD AFFORD TO BUY BOOKS FOR THE FIRST TIME EVER!

I LIKE TO MAKE A "DOUBLE" BY LOOKING IN THE MIRROR. THERE ARE 2 OF ME--IT'S LIKE I MADE A *COPY* OF MYSELF!

Remember the *doubles*, from when you learned your addition facts? (See p. 41 in *Do Not Open This Math Book* for more on this!) They're going to help us a lot—and we'll add two more facts to the list: 11 + 11 = 22 and 12 + 12 = 24. Not too bad, right?

1 + 1 = 2	2 + 2 = 4	3 + 3 = 6	4 + 4 = 8	5 + 5 = 10	6 + 6 = 12
7 + 7 = 14	8 + 8 = 16	9 + 9 = 18	10 + 10 = 20	11 + 11 = 22	12 + 12 = 24

If we've got two 3's, we can write it with addition or multiplication:

Two Ways of Writing "Two 3's"

$$3 + 3 = 6 \qquad 2 \times 3 = 6$$

two 3's　　　　two times 3

It's the same thing, written differently! Now, doesn't the chart below look nice and friendly? Remember, when we multiply something times 2, we're just doubling it!

2 × 1 = 2	2 × 2 = 4	2 × 3 = 6	2 × 4 = 8	2 × 5 = 10	2 × 6 = 12
2 × 7 = 14	2 × 8 = 16	2 × 9 = 18	2 × 10 = 20	2 × 11 = 22	2 × 12 = 24

Here are the same 2's facts shown with their arrays and their entire fact families:

● ●	● ● ● ●	● ● ● ● ● ●	● ● ● ● ● ● ● ●	● ● ● ● ● ● ● ● ● ●
2 × 1 = 2	 2 × 2 = 4	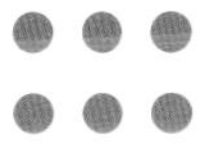 2 × 3 = 6	 2 × 4 = 8	 2 × 5 = 10
1 × 2 = 2	4 ÷ 2 = 2	3 × 2 = 6	4 × 2 = 8	5 × 2 = 10
2 ÷ 1 = 2	(It's a square number!)	6 ÷ 3 = 2	8 ÷ 4 = 2	10 ÷ 5 = 2
2 ÷ 2 = 1		6 ÷ 2 = 3	8 ÷ 2 = 4	10 ÷ 2 = 5

See the next page for more!

●●●●●● ●●●●●●	●●●●●●● ●●●●●●●	●●●●●●●● ●●●●●●●●	●●●●●●●●● ●●●●●●●●●
2 × 6 = 12	2 × 7 = 14	2 × 8 = 16	2 × 9 = 18
6 × 2 = 12	7 × 2 = 14	8 × 2 = 16	9 × 2 = 18
12 ÷ 6 = 2	14 ÷ 7 = 2	16 ÷ 8 = 2	18 ÷ 9 = 2
12 ÷ 2 = 6	14 ÷ 2 = 7	16 ÷ 2 = 8	18 ÷ 2 = 9

●●●●●●●●●● ●●●●●●●●●●	●●●●●●●●●●● ●●●●●●●●●●●	●●●●●●●●●●●● ●●●●●●●●●●●●
2 × 10 = 20	2 × 11 = 22	2 × 12 = 24
10 × 2 = 20	11 × 2 = 22	12 × 2 = 24
20 ÷ 10 = 2	22 ÷ 11 = 2	24 ÷ 12 = 2
20 ÷ 2 = 10	22 ÷ 2 = 11	24 ÷ 2 = 12

It looks like a lot of information, but it's just doubles—stuff you've already learned! Let's practice our facts for the 0's, 1's, and 2's. . . .

GAME TIME!

Time to practice the 0's, 1's, and 2's multiplication facts!
I'll do the first one for you.

1. $9 \times 2 =$ _?_

Let's Play: Hmm, since multiplying by 2 is just doubling a number, we can remember from our addition facts that $9 + 9 = 18$, which means $9 \times 2 = 18$, too!

Answer: $9 \times 2 = 18$

2. $2 \times 2 =$ _?_   3. $5 \times 0 =$ _?_   4. $1 \times 6 =$ _?_   5. $5 \times 2 =$ _?_

6. $11 \times 2 =$ _?_   7. $8 \times 1 =$ _?_   8. $10 \times 2 =$ _?_   9. $2 \times 8 =$ _?_

10. $2 \times 3 =$ _?_   11. $6 \times 2 =$ _?_   12. $0 \times 1 =$ _?_   13. $8 \times 2 =$ _?_

14. $2 \times 4 =$ _?_   15. $8 \times 0 =$ _?_   16. $2 \times 9 =$ _?_   17. $7 \times 2 =$ _?_

18. $1 \times 1 =$ _?_   19. $12 \times 1 =$ _?_   20. $2 \times 12 =$ _?_   21. $0 \times 0 =$ _?_

I recommend doing those a few times! Also find more practice at **TheTimesMachine.com**. Then, <u>once you're pretty comfortable with them</u>, try these division problems. I'll do the first one for you!

Keep going! $\longrightarrow$

(Answers on page 221.)

1. 14 ÷ 2 = ___?___

Let's Play: Time to stand on our heads and rethink this division problem as multiplication with something missing! So it becomes 2 x ? = 14. Hmm . . . this looks familiar! In #17 on the previous page, we did 7 x 2 = 14. And because of fact families, that means 2 x 7 = 14 is also true. So 7 was the missing number! And that means 14 ÷ 2 = 7. Done!

Answer: 14 ÷ 2 = 7

2. 2 ÷ 1 = ___?___

3. 0 ÷ 5 = ___?___

4. 6 ÷ 6 = ___?___

5. 10 ÷ 2 = ___?___

6. 22 ÷ 11 = ___?___

7. 8 ÷ 1 = ___?___

8. 20 ÷ 2 = ___?___

9. 16 ÷ 2 = ___?___

10. 6 ÷ 3 = ___?___

11. 12 ÷ 2 = ___?___

12. 0 ÷ 1 = ___?___

13. 16 ÷ 8 = ___?___

14. 8 ÷ 2 = ___?___

15. 12 ÷ 1 = ___?___

16. 18 ÷ 2 = ___?___

17. 14 ÷ 7 = ___?___

18. 1 ÷ 1 = ___?___

19. 12 ÷ 12 = ___?___

20. 24 ÷ 12 = ___?___

21. 18 ÷ 9 = ___?___

Great job! Now I suggest spending at least a week or two memorizing these before moving on to the 3's. The best way to learn them? Read the 2's section every day for a week, and practice these problems every day, too. Your parents can also download and print multiplication facts from **TheTimesMachine.com** to tape on your bathroom mirror, so twice a day you'll see them when you're brushing your teeth! Then, once you feel like you know them really well, move on to the 3's. You got this!

(Answers on page 221.)

The THREES . . . and Beyond!

Now that we're getting into the 3's, the numbers will start to get a little bigger, so we'll spend more time on each one! For some of these next facts, the Times Machine has a little story to help you remember the answers. First, let's get to know the cute animals and objects that will be in these little stories!

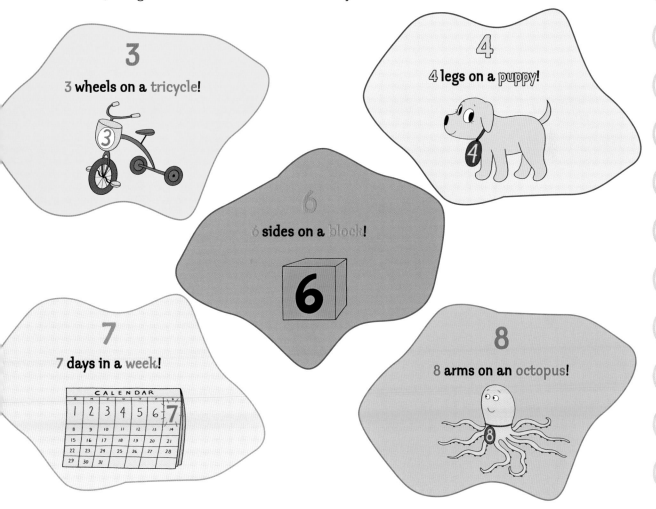

So if you want to remember 4 × 6, you can just think about the story of the 4-legged puppy who climbed up on the 6-sided block, or if you want to remember 8 × 3, just think about the story of the 8-armed octopus who tried to ride a 3-wheeled tricycle!

There's no 5, 9, 10, 11, or 12, because those have their own tricks—no stories needed!

I'll show you lots of helpful tips, tricks, and stories for memorizing these facts. We'll learn (and practice) multiplication and division facts up through 8 × 8 before moving on to the 9's–12's, which have their own fun tricks. **Use whichever reminders and tips are most helpful to YOU.** Let's begin!

THE THREES

On p. 75, we saw that to multiply by 2, we can simply double the number to get our answer, like how 2 × 4 is the same as 4 + 4 = 8. It's our doubles! Well, for the 3's, we can just double the number, and then add *one more* of that number! So to multiply 3 × 4, since we know 2 × 4 = 8, we just add *one more 4* to get **12**!

3 × 4 = ?   That means we need three 4's!

2 × 4 = 8 ◄——— We can double 4 to get two 4's!

   + 4 ◄——— Here's one more 4.

= 12   So this is three 4's, our answer!

Ta-da! 3 × 4 = **12**

You can do this with any of our 3's multiplication facts, but I'll give you other ways to remember them, too. Use whichever ones you like the most. Let's do it!

3 × 1 = 3	3 × 2 = 6
(See p. 73 for this one!)	(See p. 75 for this one!)

Once upon a time, two friends used to ride their 3-wheeled tricycles every day after school to play tic-tac-toe together.

Have you ever played tic-tac-toe? It's 3 in every direction, and you guessed it—9 spots total!

3 × 3 = 9

(It's a square number!)

Fact Family

3 × 3 = 9

9 ÷ 3 = 3

Tic tac toe? Victory's *mine*!
*Three **times** three **is equal to** nine!*

OH, I LOVE SQUARE NUMBERS! THIS 3-BY-3 SQUARE LOOKS LIKE A TIC-TAC-TOE BOARD, OR LIKE WHERE YOU DIAL NUMBERS ON A PHONE!

EXCEPT NOT THE ZERO PART, DUH. JUST THE DIGITS 1 TO 9 ON A PHONE.

RIGHT--BECAUSE THERE ARE 9 OF THEM, AND 3 X 3 = 9!

Once upon a time, a 4-legged puppy really wanted to go for a ride in the 3-wheeled tricycle down 12th Street, so she climbed into the basket when her owner wasn't looking!

3 × 4 = 12

Fact Family

3 × 4 = 12

4 × 3 = 12

12 ÷ 3 = 4

12 ÷ 4 = 3

Puppy climbed up all by *herself*!
*Three **times** four **is equal to** twelf.*

WAIT--"TWELF"?!

YOU TRY RHYMING SOMETHING WITH "TWELVE"!

3 × 5 = 15

Just skip count by 5's a total of 3 times: 5, 10, **15**.

See p. 92 for more!

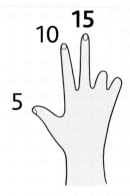

Fact Family

3 × 5 = 15
5 × 3 = 15
15 ÷ 3 = 5
15 ÷ 5 = 3

3 × 6 = 18

Once upon a time, a 3-wheeled tricycle tried to get back into its toy chest, and the queen doll helped him get in with a 6-sided block. She was a really great queen and so beautiful, too! She had 18 little gems on her coat, which were also great. Did we mention how *great* she was?!

Fact Family

3 × 6 = 18
6 × 3 = 18
18 ÷ 3 = 6
18 ÷ 6 = 3

"Wow! You're such a *great queen!*"
Three **times** **six** **is eighteen.**

$3 \times 7 = 21$

Once upon a time, when Ms. Squirrel was little, she spent an entire week during the summer riding her 3-wheeled tricycle—yep, every day for 7 days! It was really hot outside and she got pretty <u>sweaty</u>, but she didn't mind because it was so much <u>fun</u>!

Fact Family

$3 \times 7 = 21$

$7 \times 3 = 21$

$21 \div 3 = 7$

$21 \div 7 = 3$

> **Ride in the sun? That's *sweaty fun*!**
> **Three *times seven is twenty-one*.**

$3 \times 8 = 24$

Have you ever seen an 8-armed octopus ride a 3-wheeled tricycle? It doesn't work very well! When Mr. Octopus tried (and he tried ALL day and night—24 hours!), all his extra arms kept getting tangled—and the <u>more</u> he tried, the <u>more</u> tangled they got!

Fact Family

$3 \times 8 = 24$

$8 \times 3 = 24$

$24 \div 3 = 8$

$24 \div 8 = 3$

A trick for multiplying by 8 is *doubling 4:*
Since $3 \times \underline{4} = 12$, just double the 12 and we get $3 \times \underline{8} = 24$.

> **His arms got tangled, *more and more*.**
> **Three *times eight is twenty-four*.**

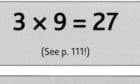

3 × 9 = 27
(See p. 111!)

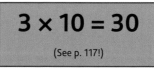

3 × 10 = 30
(See p. 117!)

3 × 11 = 33
(See p. 123!)

3 × 12 = 36
(See p. 130!)

The Adding-the-Digits Trick

For any of the 3's facts, we can add up the digits of the product . . . and get something else divisible by 3! For example, 3 × 7 = 21, and if we add up the digits of 21, we get 2 + 1 = **3**. How about that! Also, 3 × 12 = 36, and if we add up the digits of 36, we get 3 + 6 = **9**, which is of course also divisible by 3. Neat, right? *This ONLY works for 3, though.* (But there's a similar trick for the 9's—see p. 109!) It's a great way to help check our answers!

WHOA...!

Be sure to <u>learn the stories and tricks</u> in this 3's section *first*.
After that, it's time to practice the 3's multiplication facts: 3 × 0 to 3 × 8, the ones we've learned so far. The colorful problems will help you remember their stories, and the rest use other tricks we've learned.
I'll do the first one for you!

1. 8 × 3 = _?_

Let's Play: Hmm, we see an *8,* which means Mr. Octopus, and a *3,* which means the tricycle . . . oh right! Mr. Octopus tried to ride the tricycle but all his arms got in the way, and his arms got tangled, more and more, which rhymes with twenty-four. And hey, 2 + 4 = 6, which, yep, is divisible by *3!* (See the trick on p. 84!)

Answer: *8 × 3 = 24*

2. 3 × 3 = _?_

3. 3 × 6 = _?_

4. 3 × 4 = _?_

5. 3 × 8 = _?_

6. 3 × 7 = _?_

7. 4 × 3 = _?_

8. 5 × 3 = _?_

9. 3 × 0 = _?_

10. 3 × 2 = _?_

11. 8 × 3 = _?_

12. 3 × 5 = _?_

13. 7 × 3 = _?_

14. 2 × 3 = _?_

15. 5 × 3 = _?_

16. 1 × 3 = _?_

17. 6 × 3 = _?_

I recommend doing those a few times! Also find more practice at

TheTimesMachine.com. Then, <u>once you're pretty comfortable with them</u>, try the

division problems on the next page. I'll do the first one for you!

Keep going! ⟶

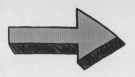

GAME TIME!

1. $21 \div 7 =$ __?__

Let's Play: This might look hard at first, but we'll just stand on our heads and rethink this division problem as multiplication with something missing! So it becomes 7 x ? = 21. Hmm . . . this looks familiar! In #13 on the previous page, we did 7 x **3** = 21. So **3** was the missing number! And that means 21 ÷ 7 = **3**. Done!

Answer: 21 ÷ 7 = **3**

2. $6 \div 2 =$ __?__   3. $15 \div 3 =$ __?__   4. $18 \div 3 =$ __?__   5. $6 \div 3 =$ __?__

6. $15 \div 5 =$ __?__   7. $0 \div 3 =$ __?__   8. $21 \div 7 =$ __?__   9. $24 \div 3 =$ __?__

10. $9 \div 3 =$ __?__   11. $3 \div 1 =$ __?__   12. $21 \div 3 =$ __?__   13. $12 \div 4 =$ __?__

14. $24 \div 8 =$ __?__   15. $12 \div 3 =$ __?__   16. $18 \div 6 =$ __?__

Great job! Read the 3's section every day for a week, and practice these problems every day for a week, too. You can also go to **TheTimesMachine.com** to download and print multiplication facts to tape on your bathroom mirror, so twice a day you'll see them when you're brushing your teeth! Then, once you feel like you know them really well, move on to the 4's. I'm so proud of you!

(Answers on page 221.)

86

× ~~~~ # THE FOURS ~~~~ ×

Remember, 4 is just double 2. So if you know **2** × a number, then just double the answer for **4** × that number! For example, since we know that 2 × 6 = **12**, to figure out <u>4</u> × 6, we can just <u>double</u> 12, and now we know that 4 × 6 = **24**. I'll also tell you some stories in this section with a yellow 4-legged puppy to help you remember some of these 4's facts. Use whichever methods help you the most!

4 × 1 = 4	4 × 2 = 8	4 × 3 = 12
(See p. 73 for this one!)	(See p. 75 for this one!)	(See p. 81 for this one!)

Once upon a time, there were two of the cutest 4-legged puppies ever, and they loved to lick everything. They would even go out to the yard and lick sticks. One day they found 16 sticks in the shape of a square. The sticks were dirty, but the puppies licked them until the sticks were *clean*!

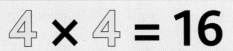

(It's a square number!)

Fact Family

4 × 4 = 16

16 ÷ 4 = 4

Also, since 2 × 4 = **8**, we know **4 × 4 = 16**, since 16 is just double 8!

*Puppies licked the **sticks clean**!*
*Four **times** four is sixteen.*

Just skip count by 5's a total of 4 times (5, 10, 15, **20**)! See p. 93 for more!

4 × 5 = 20

Once upon a time, a cute 4-legged puppy tried to stand on a small 6-sided block, but—oops! She tumbled down to the floor! Ugh! She felt so silly falling from the <u>block</u> to the <u>floor</u>!

$$4 \times 6 = 24$$

AND REMEMBER, 4 IS JUST DOUBLING A DOUBLE. SO IF YOU KNOW *2* X 6 = *12*, JUST DOUBLE THE ANSWER TO GET *4* X 6 = *24*.

Fact Family

4 × 6 = 24

6 × 4 = 24

24 ÷ 4 = 6

24 ÷ 6 = 4

The puppy fell from the block to the floor.
Four **times** *six* **is twenty-four.**

Once upon a time, there was a boy who took his cute 4-legged puppy on a walk at 4:28 every day after school and on weekends. Yep, 7 days a week. This was great, but the only problem was that the puppy liked to RUN the whole time, and the boy got tired doing that every day. So he decided to use roller skates and get skates for the puppy, too!

First we ran, and *now we skate!*
Four **times** *seven* **is twenty-eight.**

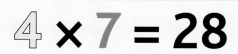

$$4 \times 7 = 28$$

Fact Family

4 × 7 = 28

7 × 4 = 28

28 ÷ 4 = 7

28 ÷ 7 = 4

And remember, 4 is just doubling a double. So if you know 2 × 7 = 14, just double the answer to get 4 × 7 = 28.

$$4 \times 8 = 32$$

Once upon a time, a cute 4-legged puppy chased a ball into the water, and she got stuck in some mud for over half an hour— 32 minutes! Mr. Octopus used his 8 arms to free the puppy and push her onto the shore. Then they both laughed because of how muddy and dirty they got!

Fact Family

$4 \times 8 = 32$

$8 \times 4 = 32$

$32 \div 4 = 8$

$32 \div 8 = 4$

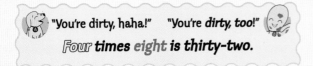

"You're dirty, haha!" "You're *dirty, too!*"
Four **times** *eight* **is thirty-two.**

Or we could double a 2's fact: Since $8 \times 2 = 16$, just *double* that for 8×4. And what's double 16? 32!

$4 \times 9 = 36$
(See p. 111!)

$4 \times 10 = 40$
(See p. 117!)

$4 \times 11 = 44$
(See p. 123!)

$4 \times 12 = 48$
(See p. 130!)

The Core of the Times Machine   **89**

 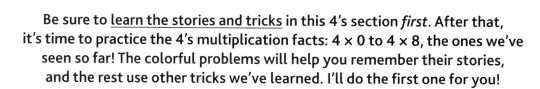 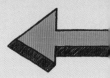

GAME TIME!

Be sure to <u>learn the stories and tricks</u> in this 4's section *first*. After that, it's time to practice the 4's multiplication facts: 4 × 0 to 4 × 8, the ones we've seen so far! The colorful problems will help you remember their stories, and the rest use other tricks we've learned. I'll do the first one for you!

1. 8 × 4 = __?__

Let's Play: Hmm, no easy numbers like 2 or 5, but we see an *8*, which means Mr. Octopus, and a *4*, which means the puppy . . . and oh yeah, Mr. Octopus helped the puppy get unstuck from the mud! And they said, "You're dirty, haha!" "You're dirty, too!" which rhymes with "thirty-two." We could also notice that *8 x 4* is just *double 8 x 2*. And what's *double 16? 32!*

Answer: 8 x 4 = 32

2. 4 × 4 = __?__   3. 4 × 7 = __?__   4. 4 × 6 = __?__   5. 4 × 8 = __?__

6. 4 × 3 = __?__   7. 2 × 4 = __?__   8. 1 × 4 = __?__   9. 4 × 0 = __?__

10. 5 × 4 = __?__   11. 4 × 1 = __?__   12. 4 × 5 = __?__   13. 3 × 4 = __?__

14. 8 × 4 = __?__   15. 4 × 2 = __?__   16. 6 × 4 = __?__   17. 7 × 4 = __?__

I recommend doing those a few times! Also find more practice at **TheTimesMachine.com**. Then, <u>once you're pretty comfortable with them</u>, try the division problems on the next page. I'll do the first one for you!

(Answers on page 221.)

1. $24 \div 4 = $ __?__

Let's Play: Time to stand on our heads and rethink this division problem as multiplication with something missing! So it becomes 4 X ? = 24. Hmm . . . this looks familiar! In #4 on the previous page, we did 4 X 6 = 24. So 6 was the missing number! And that means 24 ÷ 4 = 6. Done!

Answer: 24 ÷ 4 = 6

2. $16 \div 4 = $ __?__

3. $20 \div 4 = $ __?__

4. $24 \div 6 = $ __?__

5. $4 \div 2 = $ __?__

6. $12 \div 3 = $ __?__

7. $4 \div 4 = $ __?__

8. $0 \div 4 = $ __?__

9. $32 \div 4 = $ __?__

10. $12 \div 4 = $ __?__

11. $20 \div 5 = $ __?__

12. $28 \div 7 = $ __?__

13. $24 \div 4 = $ __?__

14. $28 \div 4 = $ __?__

15. $4 \div 1 = $ __?__

16. $8 \div 2 = $ __?__

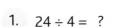

Great job! Read the 4's section every day for a week, and practice these problems every day for a week, too. You can also go to **TheTimesMachine.com** to download and print multiplication facts to tape on your bathroom mirror, so twice a day you'll see them when you're brushing your teeth! Then, once you feel like you know them really well, move on to the 5's. I'm so proud of you!

THE FiVES

I'M BEGINNING TO THINK MY FAVORITE NUMBER IS 5. IT REMINDS ME OF GIVING HIGH-FIVES, WHICH IS ONE OF MY FAVORITE THINGS TO DO! AND I LOVE SKIP-COUNTING BY 5'S. CAN WE DO THAT?

WE SURE CAN!

To figure out 5 × 4, we could skip count by 5's for each finger, and use 4 fingers!

"5!"

1st finger

5 × 1 = **5**

"10!"

2nd finger

5 × 2 = **10**

"15!"

3rd finger

5 × 3 = **15**

"20!"

4th finger

5 × 4 = **20**

5 × 1 = 5	5 × 2 = 10
(See p. 73 for this one!)	(See p. 75 for this one!)

5 × 3 = 15

Skip count by 5's

three times to get **15**:

10 15

5

Fact Family

5 × 3 = 15

3 × 5 = 15

15 ÷ 5 = 3

15 ÷ 3 = 5

The Half-of-Ten Trick

And here's another trick for 5's: Since 5 is half of 10, try multiplying a number by 10 instead of 5, and then cut the result in half for the answer! For example, to solve 5 × 8, we could first do 10 × 8 = 80, and then divide by 2 to get our answer: 40. So, 5 × 8 = 40. Ta-da!

WHOA . . . !

MULTIPLYING BY 10 IS AWESOME. AND DIVIDING BY 2 ISN'T VERY HARD. SO THIS TRICK GIVES US THE RIGHT ANSWER BECAUSE MULTIPLYING BY 10 AND THEN DIVIDING BY 2 IS THE SAME AS MULTIPLYING BY 5?

YOU GOT IT!

5 × 4 = 20

10 15 20

5

Skip count by 5's four times to get **20**:

OR

Multiply *10 × 4 = 40* and then divide in *half* to get **20**!

Fact Family

5 × 4 = 20

4 × 5 = 20

20 ÷ 5 = 4

20 ÷ 4 = 5

5 × 5 = 25

(It's a square number!)

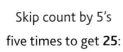

10 15 20 25

5

Skip count by 5's five times to get **25**:

OR

Multiply *10 × 5 = 50* and then divide in *half* to get **25**!

Fact Family

5 × 5 = 25

25 ÷ 5 = 5

5 × 6 = 30

Skip count by 5's
six times to get **30**:

10 15 20
5   25 **30**

OR

Multiply *10* × 6 = 60 and
then divide in *half* to get **30**!

I JUST NOTICED THAT EVERY MULTIPLE OF 5 ENDS IN EITHER A 5 OR A O. IS THAT RIGHT?

YEP--IF WE MULTIPLY 5 TIMES ANY ODD NUMBER, THE ANSWER WILL END IN 5. IF WE MULTIPLY 5 TIMES ANY EVEN NUMBER, THE ANSWER WILL END IN O!

5 × 7 = 35

Skip count by 5's
seven times to get **35**:

10 15 20 **35**
5   25 30

OR

Multiply *10* × 7 = 70
and divide in *half* to get **35**!

Fact Family

5 × 7 = 35

7 × 5 = 35

35 ÷ 5 = 7

35 ÷ 7 = 5

5 × 8 = 40

Skip count by 5's
eight times to get **40**:

10 15 20 35 **40**
5   25 30

OR

Multiply *10* × 8 = 80
and divide in *half* to get **40**!

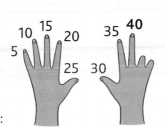

Fact Family

5 × 8 = 40

8 × 5 = 40

40 ÷ 5 = 8

40 ÷ 8 = 5

$5 \times 9 = 45$	$5 \times 10 = 50$	$5 \times 11 = 55$	$5 \times 12 = 60$
(See p. 112!)	(See p. 118!)	(See p. 124!)	(See p. 130!)

SO THESE 5'S AREN'T SO BAD, YOU KNOW? BUT 10'S ARE BETTER, SO I REALLY LIKE THAT HALF-OF-TEN TRICK, ESPECIALLY FOR BIGGER FACTS LIKE 5 X 7 AND 5 X 8.

WELL, I LIKE SKIP COUNTING-- I'M GONNA HIGH-FIVE MY WAY TO AN A+ IN MATH CLASS!

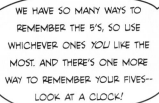

WE HAVE SO MANY WAYS TO REMEMBER THE 5'S, SO USE WHICHEVER ONES *YOU* LIKE THE MOST. AND THERE'S ONE MORE WAY TO REMEMBER YOUR FIVES-- LOOK AT A CLOCK!

The Fives "Clock" Trick!

On most clocks that use "hands" to show time, each number tells us what hour it is when we look at the short hand, but when we look at the *long* hand, each number also tells us which *minute* we're on. How? Each number means *5 minutes more* than the previous number. For example, if we look at the clock and see the short hand pointing past the 2 and the big hand pointing to the <u>3</u>, then we know that it's 2:**15**. And guess what? <u>3</u> $\times$ 5 = **15**! If instead the big hand points straight down to the <u>6</u>, then we know it's 2:**30**, and yep, <u>6</u> $\times$ 5 = **30**. Pretty cool, right?

WHOA . . . !

2:15

2:30

 # GAME TIME!

Time to practice the 5's multiplication facts: 5 × 0 to 5 × 8, the ones we've learned so far. To get the answers (until they are memorized!), we can either skip count by 5's, or multiply by 10 and then divide the answer in half. I'll do the first one for you.

1. 7 × 5 = _?_

Let's Play: We can either skip count by 5's seven times: 5, 10, 15, 20, 25, 30, **35**. Or we can multiply 7 x 10 = 70, and then divide that answer in half for **35**. Done!

Answer: 7 X 5 = 35

2. 5 × 4 = _?_   3. 1 × 5 = _?_   4. 5 × 6 = _?_   5. 5 × 3 = _?_

6. 7 × 5 = _?_   7. 0 × 5 = _?_   8. 8 × 5 = _?_   9. 5 × 5 = _?_

10. 2 × 5 = _?_   11. 6 × 5 = _?_   12. 3 × 5 = _?_   13. 4 × 5 = _?_

14. 5 × 8 = _?_   15. 5 × 2 = _?_   16. 5 × 7 = _?_   17. 5 × 0 = _?_

I recommend doing those a few times! Also find more practice at **TheTimesMachine.com**. Then, <u>once you're pretty comfortable with them</u>, try the division problems on the next page. I'll do the first one for you!

THE 5'S ARE GREAT!

I AGREE!

(Answers on page 221.)

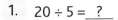

1. $20 \div 5 =$ ___?___

Let's Play: Time to stand on our heads and rethink this division problem as multiplication with something missing! So it becomes 5 x ? = 20, or "5 times *what* equals 20?" Hmm . . . this looks familiar! In #2 on the previous page, we did 5 x 4 = 20. So 4 was the missing number! And that means 20 ÷ 5 = 4. Done!

Answer: 20 ÷ 5 = 4

2. $50 \div 5 =$ ___?___   3. $25 \div 5 =$ ___?___   4. $30 \div 5 =$ ___?___   5. $10 \div 5 =$ ___?___

6. $15 \div 5 =$ ___?___   7. $40 \div 8 =$ ___?___   8. $20 \div 4 =$ ___?___   9. $40 \div 5 =$ ___?___

10. $15 \div 3 =$ ___?___   11. $30 \div 6 =$ ___?___   12. $35 \div 5 =$ ___?___   13. $0 \div 5 =$ ___?___

14. $35 \div 7 =$ ___?___   15. $5 \div 1 =$ ___?___   16. $45 \div 5 =$ ___?___   17. $10 \div 2 =$ ___?___

Great job! Read the 5's section every day for a week, and practice these problems every day for a week, too. You can also go to **TheTimesMachine.com** to download and print multiplication facts to tape on your bathroom mirror, so twice a day you'll see them when you're brushing your teeth! Then, once you feel like you know them really well, move on to the 6's. I'm so proud of you!

(Answers on page 221.)

THE SIXES

Let's jump right in!

6 × 1 = 6 (See p. 73!)	**6 × 2 = 12** (See p. 76!)

6 × 3 = 18 (See p. 82!)	**6 × 4 = 24** (See p. 88!)	**6 × 5 = 30** (See p. 94!)

Once upon a time, a little girl decided to make a big house for the squirrels that came into her backyard. She used her 6-sided blocks as the "bricks" to build it, which was great . . . but the blocks got really dirty from being outside! Look at those dirty bricks!

$$6 \times 6 = 36$$

(It's a square number!)

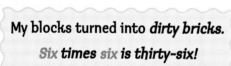

My blocks turned into *dirty bricks.*
*Six **times** six is thirty-six!*

Fact Family

6 × 6 = 36

36 ÷ 6 = 6

HEY, I KNOW THAT GUY!

AND THE RHYTHM IS GREAT--SAY IT OUT LOUD: "SIX TIMES SIX IS THIRTY-SIX!" THIS IS ONE OF MY FAVORITES BECAUSE OF HOW IT SOUNDS.

Once upon a time, a little mouse ate a HUGE 6-sided block of cheese every day for an entire week—yep, 7 days in a row! He got really full, and let's face it, he ended up farting a few times from all that cheese . . . 42 farts, in fact!

6 × 7 = 42

**The mouse got full and *farty, too!*
Six times seven is forty-two!**

Fact Family

6 × 7 = 42

7 × 6 = 42

42 ÷ 6 = 7

42 ÷ 7 = 6

Once upon a time, an 8-armed octopus was swimming in the ocean and playing with his 6-sided floating cube. But the floaty's plug came out, and it started to deflate. When the octopus used its suction cups to pull the floaty's walls back out, some water—and a tiny fish!—got sucked into the hole. That made it look like the floaty toy ATE the fish! What?! The floaty *ate something*?!

6 × 8 = 48

**Look at what the *floaty ate!*
Six times eight is forty-eight.**

 And don't worry, the fish got out safely!

Fact Family

6 × 8 = 48

8 × 6 = 48

48 ÷ 6 = 8

48 ÷ 8 = 6

---

6 × 9 = 54	**6 × 10 = 60**	**6 × 11 = 66**	**6 × 12 = 72**
(See p. 113)	(See p. 118!)	(See p. 124!)	(See p. 131!)

The Core of the Times Machine ⭐ 99

GAME TIME!

Be sure to <u>learn the stories and tricks</u> in this 6's section *first*. After that, it's time to practice the 6's multiplication facts: 6 × 0 to 6 × 8, the ones we've seen so far. The colorful problems will help you remember their stories, and the rest use other tricks we've learned. I'll do the first one for you!

1. 7 × 6 = _?_

Let's Play: Hmm, there are no "easy" numbers like 2 or 5, so that means there's a story. We see a 6, which means a 6-sided block, and a 7, which means "every day of the week" . . . which story was that? Oh, yeah! Mr. Mouse ate a huge 6-sided block of cheese every day of the week, which made him full and "farty, too!"—which sounds like "forty-two"! Yep, 7 × 6 = 42. We could also notice that 7 × 6 is one more 7 than 7 × 5, which is 35, and 35 + 7 = 42. Done!

Answer: 7 × 6 = 42

2. 6 × 4 = _?_   3. 6 × 8 = _?_   4. 6 × 3 = _?_   5. 6 × 7 = _?_

6. 6 × 6 = _?_   7. 3 × 6 = _?_   8. 1 × 6 = _?_   9. 6 × 2 = _?_

10. 6 × 0 = _?_   11. 7 × 6 = _?_   12. 6 × 5 = _?_   13. 8 × 6 = _?_

14. 6 × 1 = _?_   15. 2 × 6 = _?_   16. 5 × 6 = _?_   17. 4 × 6 = _?_

I recommend doing those a few times! Also find more practice at **TheTimesMachine.com**. Then, <u>once you're pretty comfortable with them</u>, try the division problems on the next page. I'll do the first one for you!

(Answers on page 221.)

1. $48 \div 6 = \underline{\ ?\ }$

Let's Play: Time to stand on our heads and rethink this division problem as multiplication with something missing! So it becomes $6 \times ? = 48$. Hmm . . . this looks familiar! In #3 on the previous page, we did $6 \times 8 = 48$. So 8 was the missing number! And that means $48 \div 6 = 8$. Done!

Answer: $48 \div 6 = 8$

2. $24 \div 4 = \underline{\ ?\ }$

3. $30 \div 6 = \underline{\ ?\ }$

4. $24 \div 6 = \underline{\ ?\ }$

5. $12 \div 2 = \underline{\ ?\ }$

6. $18 \div 3 = \underline{\ ?\ }$

7. $48 \div 8 = \underline{\ ?\ }$

8. $6 \div 6 = \underline{\ ?\ }$

9. $48 \div 6 = \underline{\ ?\ }$

10. $18 \div 6 = \underline{\ ?\ }$

11. $0 \div 6 = \underline{\ ?\ }$

12. $42 \div 6 = \underline{\ ?\ }$

13. $6 \div 1 = \underline{\ ?\ }$

14. $42 \div 7 = \underline{\ ?\ }$

15. $48 \div 6 = \underline{\ ?\ }$

16. $12 \div 6 = \underline{\ ?\ }$

Great job! Read the 6's section every day for a week, and practice these problems every day for a week, too. You can also go to **TheTimesMachine.com** to download and print multiplication facts to tape on your bathroom mirror, so twice a day you'll see them when you're brushing your teeth! Then, once you feel like you know them really well, move on to the 7's. I'm so proud of you!

(Answers on page 221.)

THE SEVENS

Let's do this!

7 × 1 = 7 (See p. 73!)	**7 × 2 = 14** (See p. 76!)	**7 × 3 = 21** (See p. 83!)
7 × 4 = 28 (See p. 88!)	**7 × 5 = 35** (See p. 94!)	**7 × 6 = 42** (See p. 99!)

Do you like sports? Have you ever seen snow skis and poles? Look at Ms. Squirrel—she's ready to hit the slopes! And if you use your imagination, you'll see that her ski poles look sort of like the blue multiplication symbol, and her skis look a little like our long green 7's, but pushed over on their sides! Hey, that's kind of like 7 × 7!

Well, Ms. Squirrel loves to ski, but she also loves *all* sports. So she made a sign with a bunch of sports on it. That's a really sporty sign, isn't it?

7 × 7 = 49

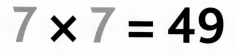

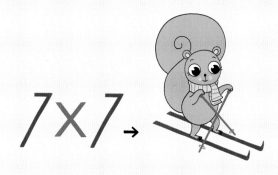

(It's a square number!)

Fact Family

7 × 7 = 49

49 ÷ 7 = 7

That's a really *sporty sign*!
***Seven* times *seven* is forty-nine.**

Once upon a time, Mr. Octopus tried to memorize the multiplication fact 7 × 8 = 56. He tried and tried, every day for a whole week—yep, 7 days! But he just couldn't remember it. Then Danica showed him a trick:

7 × 8 = 56

```
7 × 8 = 56
     ↶↷
Switch the two sides of a fact,
   and we still have a fact!
    56 = 7 × 8
```

Fact Family

7 × 8 = 56
8 × 7 = 56
56 ÷ 7 = 8
56 ÷ 8 = 7

And lookie there—*5, 6, 7, 8*—those numbers go in order! You can say *"Five, six, seven, eight!"* to help remember it. Nice, right?

5, 6, 7, 8!

Change the order; *that's the fix!*
Seven *times* eight is fifty-six.

7 × 9 = 63	7 × 10 = 70	7 × 11 = 77	7 × 12 = 84
(See p. 113!)	(See p. 118!)	(See p. 124!)	(See p. 131!)

BACK ON 7 × 7 = 49, I WAS IMPRESSED YOU COULD SKI ON THAT PAIR OF 7'S. BUT I DON'T THINK YOU'RE SUPPOSED TO CROSS YOUR POLES LIKE THAT.

I CROSSED THEM TO MAKE THE MULTIPLICATION SYMBOL!

Be sure to <u>learn the stories and tricks</u> in this 7's section *first*. After that, it's time to practice the 7's multiplication facts: 7 × 0 to 7 × 8, the ones we've seen so far. The colorful problems will help you remember their stories, and the rest use other tricks we've learned. I'll do the first one for you!

1. 4 × 7 = _?_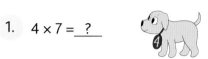

Let's Play: Hmm, there are no "easy" factors like 2 or 5, so that means there's a story. We see a 4, which means the puppy, and a 7, which means "every day of the week" . . . which story was that? Oh, yeah! We took the puppy for a walk every day, and she wanted to run, so we got some skates to make it easier. "First we ran, and now we skate! 4 × 7 is 28." Yep, 4 × 7 = 28. We could also notice that 4 × 7 is *double* 2 × 7, which is 14, and *double* 14 is 28. Done!

Answer: 4 x 7 = 28

2. 7 × 6 = _?_   3. 7 × 3 = _?_   4. 7 × 4 = _?_   5. 7 × 8 = _?_

6. 7 × 7 = _?_   7. 0 × 7 = _?_   8. 1 × 7 = _?_   9. 6 × 7 = _?_

10. 8 × 7 = _?_   11. 7 × 2 = _?_   12. 7 × 5 = _?_   13. 4 × 7 = _?_

14. 3 × 7 = _?_   15. 2 × 7 = _?_   16. 5 × 7 = _?_   17. 7 × 0 = _?_

I recommend doing those a few times! Also find more practice at **TheTimesMachine.com**. Then, <u>once you're pretty comfortable with them</u>, try the division problems on the next page. I'll do the first one for you!

(Answers on page 221.)

1. $49 \div 7 = \underline{\ ?\ }$

Let's Play: Time to stand on our heads and rethink this division problem as multiplication with something missing! So it becomes 7 x ? = 49. Hmm . . . this looks familiar! In #6 on the previous page, we did 7 x 7 = 49—hey, a square! So 7 was the missing number! And that means 49 ÷ 7 = 7. Done!

Answer: 49 ÷ 7 = 7

2. $28 \div 4 = \underline{\ ?\ }$ 3. $35 \div 7 = \underline{\ ?\ }$ 4. $28 \div 7 = \underline{\ ?\ }$ 5. $14 \div 2 = \underline{\ ?\ }$

6. $21 \div 3 = \underline{\ ?\ }$ 7. $63 \div 7 = \underline{\ ?\ }$ 8. $70 \div 7 = \underline{\ ?\ }$ 9. $56 \div 8 = \underline{\ ?\ }$

10. $21 \div 7 = \underline{\ ?\ }$ 11. $7 \div 7 = \underline{\ ?\ }$ 12. $42 \div 7 = \underline{\ ?\ }$ 13. $0 \div 7 = \underline{\ ?\ }$

14. $49 \div 7 = \underline{\ ?\ }$ 15. $63 \div 9 = \underline{\ ?\ }$ 16. $56 \div 7 = \underline{\ ?\ }$

Great job! Read the 7's section every day for a week, and practice these problems every day for a week, too. You can also go to **TheTimesMachine.com** to download and print multiplication facts to tape on your bathroom mirror, so twice a day you'll see them when you're brushing your teeth! Then, once you feel like you know them really well, move on to the 8's. I'm so proud of you!

(Answers on page 221.)

THE EIGHTS

Let's do these 8's!

LOOK AT ALL THESE WE'VE LEARNED SO FAR!

8 × 1 = 8 (See p. 73!)	**8 × 2 = 16** (See p. 76!)	**8 × 3 = 24** (See p. 83!)

8 × 4 = 32 (See p. 89!)	**8 × 5 = 40** (See p. 94!)	**8 × 6 = 48** (See p. 99!)	**8 × 7 = 56** (See p. 103!)

Once upon a time, there was an adventurous little octopus who wanted to leave the ocean and see what the inside of a house is like. Her little brother came along, too. When they went into the kitchen, they kept sticking to the floor! They didn't know it was their 64 suction cups—they just thought the floor was REALLY sticky, and said:

8 × 8 = 64

(It's a square number!)

Fact Family

8 × 8 = 64

64 ÷ 8 = 8

"Goodness! What a *sticky floor!* *Eight **times** eight is sixty-four.*

8 × 9 = 72 (See p. 114!)	**8 × 10 = 80** (See p. 118!)	**8 × 11 = 88** (See p. 125!)	**8 × 12 = 96** (See p. 131!)

GAME TIME!

Be sure to <u>learn the stories and tricks</u> in this 8's section *first*. After that, it's time to practice the 8's multiplication facts: 8 × 0 to 8 × 8, the ones we've seen so far. The colorful problems will help you remember their stories, and the rest use other tricks we've learned. I'll do the first one for you!

1.　8 × 8 = __?__

Let's Play: Hmm, there are no "easy" numbers like 2 or 5, so that means there's a story: There are two 8's . . . that's an octopus and another octopus. Which story was that? Oh, yeah! Because of the suction cups on their arms, the two octopus siblings got stuck to the floor—which they thought was a "sticky floor," and that sounds like "sixty-four"! Yep, 8 × 8 = 64. We could also notice that 8 × 8 is *double* 8 × 4, which is *32*, and *32 × 2 = 64*. Done!

Answer: 8 × 8 = 64

2.　8 × 4 = __?__　　3.　8 × 3 = __?__　　4.　8 × 6 = __?__　　5.　8 × 8 = __?__

6.　8 × 7 = __?__　　7.　1 × 8 = __?__　　8.　3 × 8 = __?__　　9.　7 × 8 = __?__

10.　5 × 8 = __?__　　11.　8 × 2 = __?__　　12.　8 × 5 = __?__　　13.　0 × 8 = __?__

14.　4 × 8 = __?__　　15.　2 × 8 = __?__　　16.　6 × 8 = __?__　　17.　8 × 1 = __?__

I recommend doing those a few times! Also find more practice at **TheTimesMachine.com**. Then, <u>once you're pretty comfortable with them</u>, try the division problems on the next page. I'll do the first one for you!

 Keep going! ⟶

(Answers on page 221.)

GAME TIME!

1. $0 \div 8 = \underline{\ ?\ }$

Let's Play: Time to stand on our heads and rethink this division problem as multiplication with something missing! So it becomes $8 \times ? = 0$. Well, anything times 0 is 0, so that means $8 \times 0 = 0$, too! And that means $0 \div 8 = 0$. Also, 0 divided into 8 groups is definitely going to have 0 in each group. (Notice that if it said $8 \div 0 = ?$, the answer would be that there IS no answer! See p. 70 for more.) Done!

Answer: $0 \div 8 = 0$

2. $32 \div 4 = \underline{\ ?\ }$ 3. $40 \div 8 = \underline{\ ?\ }$ 4. $32 \div 8 = \underline{\ ?\ }$ 5. $16 \div 2 = \underline{\ ?\ }$

6. $48 \div 6 = \underline{\ ?\ }$ 7. $56 \div 7 = \underline{\ ?\ }$ 8. $24 \div 3 = \underline{\ ?\ }$ 9. $56 \div 8 = \underline{\ ?\ }$

10. $24 \div 8 = \underline{\ ?\ }$ 11. $0 \div 8 = \underline{\ ?\ }$ 12. $48 \div 8 = \underline{\ ?\ }$ 13. $8 \div 8 = \underline{\ ?\ }$

14. $64 \div 8 = \underline{\ ?\ }$ 15. $40 \div 5 = \underline{\ ?\ }$ 16. $16 \div 8 = \underline{\ ?\ }$ 17. $8 \div 1 = \underline{\ ?\ }$

Great job! Read the 8's section every day for a week, and practice these problems every day for a week, too. You can also go to **TheTimesMachine.com** to download and print multiplication facts to tape on your bathroom mirror, so twice a day you'll see them when you're brushing your teeth! Then, once you feel like you know them really well, move on to the 9's. I'm so proud of you!

(Answers on page 221.)

THE NINES

Now we're ready to tackle the nines, which are very special, and that's why we've saved them until now. Check out their cool pattern—the tens digits go from 0 to 9, and the ones digits go from 9 to 0.

	Tens place	Ones place

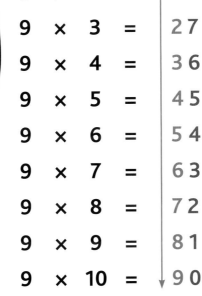

$9 \times 1 = 09$

$9 \times 2 = 18$

$9 \times 3 = 27$

$9 \times 4 = 36$

$9 \times 5 = 45$

$9 \times 6 = 54$

$9 \times 7 = 63$

$9 \times 8 = 72$

$9 \times 9 = 81$

$9 \times 10 = 90$

> WOW, THAT'S AMAZING! AND I SEE THAT WHEN A NUMBER MULTIPLIES TIMES 9, THE TENS DIGIT OF THE ANSWER IS ALWAYS 1 LESS THAN THAT NUMBER! LIKE HOW 9 X 3 = 27: THE "2" FROM 27 IS ONE LESS THAN THE "3," OR LIKE HOW 9 X 7 = 63: THE "6" FROM THE 63 IS ONE LESS THAN THE "7." AND THAT PATTERN CONTINUES!

> I LIKE THAT YOU CAN ADD UP THE DIGITS OF THE PRODUCTS AND YOU ALWAYS GET 9. LIKE FOR 9 X 6 = 54, YOU CAN ADD 5 + 4 = 9. THAT'S SORT OF REASSURING.

> GREAT JOB! BUT REMEMBER THAT THIS TRICK ONLY WORKS FOR MULTIPLES OF 9.

Another Adding-the-Digits Trick

For these 9's facts, we can add up the digits of the product . . . and get 9! For example, $9 \times 3 = 27$ and $2 + 7 = 9$. Also, $9 \times 6 = 54$ and $5 + 4 = 9$. Neat, right? This ONLY works for 9, though. (But there's a similar trick for the 3's—see p. 84!)

And remember the 9's Fingers Trick I mentioned on p. 67? I'll show you now!

The 9's Fingers Trick

For 9 times any single-digit number, we can hold our hands in front of us, bend down the number we're multiplying by, and then our fingers SHOW us the answer! For example, if we want to multiply 9 × 4, we can bend down the 4th finger, and then our fingers show us 3 and 6, which makes 36! And yep, 9 × 4 = 36. Neat, right? Below, I'll show you how this looks on fingers!

WHOA!

9 × 1 = 9
(See p. 73!)

Notice the digits of the answer should add up to 9, and yep, 1 + 8 = 9! (Remember, this adding trick ONLY works for multiples of 9.) Great!

9 × 2 = 18

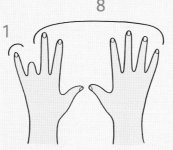

Bend down *2nd* finger for *9 × 2 = 18*

Fact Family
9 × 2 = 18
2 × 9 = 18
18 ÷ 9 = 2
18 ÷ 2 = 9

ALSO REMEMBER THE "WHAT IF THE 9 WERE A 10" TRICK: INSTEAD OF 9 X 2, WE'D HAVE 10 X 2 = 20. BUT THAT'S TOO MANY 2'S! WE DIDN'T WANT *TEN* 2'S, WE ONLY WANTED *NINE* 2'S, SO WE SUBTRACT *ONE* 2, AND GET 20 - 2 = *18*.

Notice the digits of the answer should add up to 9, and yep, 2 + 7 = 9. (Remember, this adding trick ONLY works for multiples of 9.) Nice!

$$9 \times 3 = 27$$

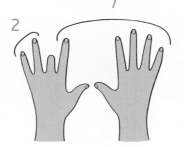

Bend down *3rd* finger for 9 × 3 = 27

ALSO REMEMBER THE "WHAT IF THE 9 WERE A 10" TRICK: INSTEAD OF 9 X 3, WE'D HAVE 10 X 3 = 30. BUT THAT'S TOO MANY 3'S! WE DIDN'T WANT *TEN* 3'S, WE ONLY WANTED *NINE* 3'S, SO WE SUBTRACT *ONE* 3, AND GET 30 - 3 = **27**.

Fact Family

9 × 3 = 27

3 × 9 = 27

27 ÷ 9 = 3

27 ÷ 3 = 9

Notice the digits of the answer should add up to 9, and yep, 3 + 6 = 9. (Remember, this adding trick ONLY works for multiples of 9.) Very nice!

$$9 \times 4 = 36$$

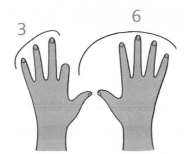

Bend down *4th* finger for 9 × 4 = 36

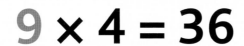

ALSO REMEMBER THE "WHAT IF THE 9 WERE A 10" TRICK: INSTEAD OF 9 X 4, WE'D HAVE 10 X 4 = 40. BUT THAT'S TOO MANY 4'S! WE DIDN'T WANT *TEN* 4'S, WE ONLY WANTED *NINE* 4'S, SO WE SUBTRACT *ONE* 4, AND GET 40 - 4 = **36**.

Fact Family

9 × 4 = 36

4 × 9 = 36

36 ÷ 9 = 4

36 ÷ 4 = 9

Notice the digits of the answer should add up to 9, and yep, 4 + 5 = 9. (Remember, this adding trick ONLY works for multiples of 9.) Great job!

9 × 5 = 45

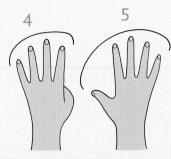

4   5

Bend down *5th* finger for 9 × 5 = 45

I FEEL SO SMART! I JUST LOVE THE 9'S. I MEAN, THE 9'S FINGER TRICK IS AMAZING!

Fact Family

9 × 5 = 45

5 × 9 = 45

45 ÷ 9 = 5

45 ÷ 5 = 9

FOR 9 X 5, WE COULD ALSO SKIP COUNT BY 5'S 9 TIMES: 5, 10, 15, 20, 25, 30, 35, 40, *45!*

ALSO REMEMBER THE "WHAT IF THE 9 WERE A 10" TRICK: INSTEAD OF 9 X 5, WE'D HAVE 10 X 5 = 50. BUT THAT'S TOO MANY 5'S! WE DIDN'T WANT *TEN* 5'S, WE ONLY WANTED *NINE* 5'S, SO WE SUBTRACT *ONE* 5, AND GET 50 - 5 = *45*.

Notice the digits of the answer should add up to 9, and yep, 5 + 4 = 9. (Remember, this adding trick ONLY works for multiples of 9.) Awesome!

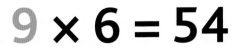

9 × 6 = 54

5   4

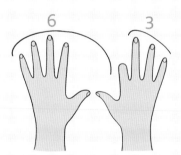

Bend down *6th* finger for 9 × 6 = 54

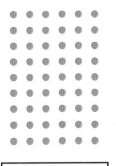

Fact Family

9 × 6 = 54

6 × 9 = 54

54 ÷ 9 = 6

54 ÷ 6 = 9

ALSO REMEMBER THE "WHAT IF THE 9 WERE A 10" TRICK: INSTEAD OF 9 X 6, WE'D HAVE 10 X 6 = 60. BUT THAT'S TOO MANY 6'S! WE DIDN'T WANT *TEN* 6'S, WE ONLY WANTED *NINE* 6'S, SO WE SUBTRACT *ONE* 6, AND GET 60 - 6 = **54**.

Notice the digits of the answer should add up to 9, and yep, 6 + 3 = 9. (Remember, this adding trick ONLY works for multiples of 9.) Fantastic!

9 × 7 = 63

6   3

Bend down *7th* finger for 9 × 7 = 63

Fact Family

9 × 7 = 63

7 × 9 = 63

63 ÷ 9 = 7

63 ÷ 7 = 9

ALSO REMEMBER THE "WHAT IF THE 9 WERE A 10" TRICK: INSTEAD OF 9 X 7, WE'D HAVE 10 X 7 = 70. BUT THAT'S TOO MANY 7'S! WE DIDN'T WANT *TEN* 7'S, WE ONLY WANTED *NINE* 7'S, SO WE SUBTRACT *ONE* 7, AND GET 70 - 7 = **63**.

Notice the digits of the answer should add up to 9, and yep, 7 + 2 = 9. (Remember, this adding trick ONLY works for multiples of 9.) Nice!

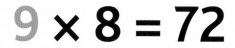

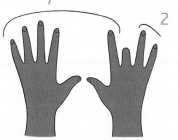

Bend down *8th* finger for 9 × 8 = 72

ALSO REMEMBER THE "WHAT IF THE 9 WERE A 10" TRICK: INSTEAD OF 9 X 8, WE'D HAVE 10 X 8 = 80. BUT THAT'S TOO MANY 8'S! WE DIDN'T WANT *TEN* 8'S, WE ONLY WANTED *NINE* 8'S, SO WE SUBTRACT *ONE* 8, AND GET 80 - 8 = *72*.

Fact Family
9 × 8 = 72
8 × 9 = 72
72 ÷ 9 = 8
72 ÷ 8 = 9

Notice the digits of the answer should add up to 9, and yep, 8 + 1 = 9. (Remember, this adding trick ONLY works for multiples of 9.) Great!

$$9 \times 9 = 81$$

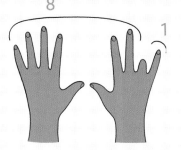

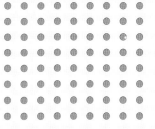

(It's a square number!)

Bend down *9th* finger for 9 × 9 = 81

ALSO REMEMBER THE "WHAT IF THE 9 WERE A 10" TRICK: INSTEAD OF 9 X 9, WE'D HAVE 10 X 9 = 90. BUT THAT'S TOO MANY 9'S! WE DIDN'T WANT *TEN* 9'S, WE ONLY WANTED *NINE* 9'S, SO WE SUBTRACT *ONE* 9, AND GET 90 - 9 = *81*.

Fact Family
9 × 9 = 81
81 ÷ 9 = 9

9 × 10 = 90
(See p. 119!)

9 × 11 = 99
(See p. 125!)

9 × 12 = 108
(See p. 132!)

GAME TIME!

Time to practice the 9's multiplication facts: 9 × 0 to 9 × 9, the ones we've learned so far. There are many tricks you can use to get the answers, and with enough practice, you'll get these memorized! I'll do the first one for you.

1. $7 \times 9 = \underline{\ ?\ }$

Let's Play: Let's do this two ways. First, what if the 9 were a 10? Then we'd get 70, right? But that's one too many 7's, so if we subtract 7, we get 70 − 7 = *63*. Or, using the 9's trick, we could hold our fingers out, put down the 7th finger, and see 6 fingers and 3 fingers: *63*. It's great to do it both ways to make sure we get the same answer. Done!

Answer: 7 x 9 = 63

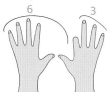

2. $3 \times 9 = \underline{\ ?\ }$ 3. $1 \times 9 = \underline{\ ?\ }$ 4. $6 \times 9 = \underline{\ ?\ }$ 5. $7 \times 9 = \underline{\ ?\ }$

6. $9 \times 2 = \underline{\ ?\ }$ 7. $9 \times 5 = \underline{\ ?\ }$ 8. $4 \times 9 = \underline{\ ?\ }$ 9. $8 \times 9 = \underline{\ ?\ }$

10. $2 \times 9 = \underline{\ ?\ }$ 11. $9 \times 9 = \underline{\ ?\ }$ 12. $5 \times 9 = \underline{\ ?\ }$ 13. $9 \times 3 = \underline{\ ?\ }$

14. $9 \times 8 = \underline{\ ?\ }$ 15. $9 \times 0 = \underline{\ ?\ }$ 16. $9 \times 7 = \underline{\ ?\ }$ 17. $9 \times 6 = \underline{\ ?\ }$

I recommend doing those a few times! Also find more practice at **TheTimesMachine.com**. Then, <u>once you're pretty comfortable with them</u>, try the division problems on the next page. I'll do the first one for you!

Keep going! ⟶

(Answers on page 221.)

GAME TIME!

1. 81 ÷ 9 = __?__

Let's Play: Time to stand on our heads and rethink this division problem as multiplication with something missing! So it becomes 9 x ? = 81. And hey, in #11 on the previous page, we did 9 x 9 = 81. And that means the missing number is 9. Done!

Answer: 81 ÷ 9 = 9

2. 36 ÷ 4 = __?__

3. 45 ÷ 9 = __?__

4. 36 ÷ 9 = __?__

5. 18 ÷ 2 = __?__

6. 72 ÷ 8 = __?__

7. 63 ÷ 7 = __?__

8. 72 ÷ 9 = __?__

9. 63 ÷ 9 = __?__

10. 0 ÷ 9 = __?__

11. 27 ÷ 9 = __?__

12. 54 ÷ 9 = __?__

13. 9 ÷ 9 = __?__

14. 81 ÷ 9 = __?__

15. 45 ÷ 5 = __?__

16. 27 ÷ 3 = __?__

17. 54 ÷ 6 = __?__

Great job! Read the 9's section every day for a week, and practice these problems every day for a week, too. You can also go to **TheTimesMachine.com** to download and print multiplication facts to tape on your bathroom mirror, so twice a day you'll see them when you're brushing your teeth! Then, once you feel like you know them really well, move on to the 10's. I'm so proud of you!

(Answers on page 221.)

THE TENS

As we saw on p. 43, to multiply a number by 10, we need to change the *place value* of the number by one, so *we can just add one zero at the end*! For example, 3 × 10 = 30, and 12 × 10 = 120. Easy, right?

To multiply by 10:

3 × 10 = 30

Just add a zero!

12 × 10 = 120

Just add a zero!

I LOVE 10!

HA! YOU SOUND LIKE ME.

10 × 1 = 10	10 × 2 = 20
(See p. 73 for this one!)	(See p. 76 for this one!)

10 × 3 = 30

Fact Family

10 × 3 = 30

3 × 10 = 30

30 ÷ 10 = 3

30 ÷ 3 = 10

10 × 4 = 40

Fact Family

10 × 4 = 40

4 × 10 = 40

40 ÷ 10 = 4

40 ÷ 4 = 10

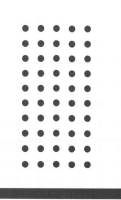

$$10 \times 5 = 50$$

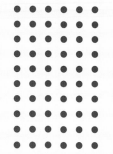

$$10 \times 6 = 60$$

Fact Family
10 × 6 = 60
6 × 10 = 60
60 ÷ 10 = 6
60 ÷ 6 = 10

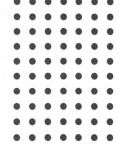

$$10 \times 7 = 70$$

Fact Family
10 × 7 = 70
7 × 10 = 70
70 ÷ 10 = 7
70 ÷ 7 = 10

$$10 \times 8 = 80$$

Fact Family
10 × 8 = 80
8 × 10 = 80
80 ÷ 10 = 8
80 ÷ 8 = 10

$$10 \times 9 = 90$$

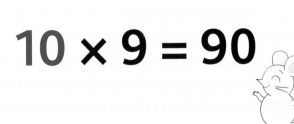

Fact Family

10 × 9 = 90

9 × 10 = 90

90 ÷ 10 = 9

90 ÷ 9 = 10

$$10 \times 10 = 100$$

Fact Family

10 × 10 = 100

100 ÷ 10 = 10

(It's a square number!)

$$10 \times 11 = 110$$

Fact Family

10 × 11 = 110

11 × 10 = 110

110 ÷ 10 = 11

110 ÷ 11 = 10

$$10 \times 12 = 120$$

Fact Family

10 × 12 = 120

12 × 10 = 120

120 ÷ 10 = 12

120 ÷ 12 = 10

If a number ends in a zero, it's really easy to divide that number by 10. We just take off a zero. After all, division undoes multiplication, like we learned about on p. 52. So if *adding a zero* means we're multiplying by 10, it makes sense that *taking off a zero* should mean we're dividing by 10! See p. 152 for more about place value and how this all works.

Examples of Multiplying and Dividing by Ten

Multiplying by 10: Add a zero!	Dividing by 10: Take off a zero!
8 × 10 = 80	80 ÷ 10 = 8
33 × 10 = 330	330 ÷ 10 = 33
456 × 10 = 4,560	4,560 ÷ 10 = 456

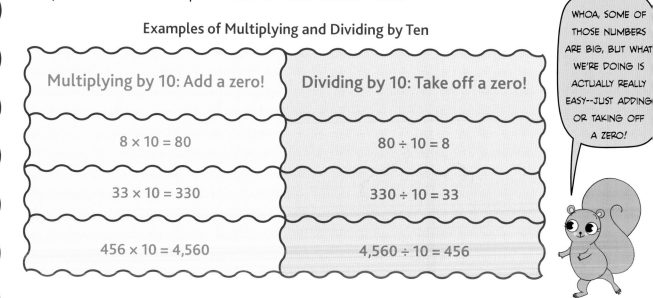

When you're ready to learn how to divide by ten when the number *doesn't* end in a zero, check out p. 122 in my book *Math Doesn't Suck*!

Time to practice multiplying by 10. These are fun because any time we multiply a number by 10, we just add a zero to it! Because of that, we'll do more than just our multiplication facts. I'll do the first one for you!

1. $34 \times 10 = $ __?__

Let's Play: To multiply 34 by 10, all we have to do is add a zero to 34, so we get 340. Done!

Answer: 34 x 10 = 340

2. $3 \times 10 = $ __?__

3. $1 \times 10 = $ __?__

4. $6 \times 10 = $ __?__

5. $11 \times 10 = $ __?__

6. $10 \times 2 = $ __?__

7. $10 \times 5 = $ __?__

8. $12 \times 10 = $ __?__

9. $8 \times 10 = $ __?__

10. $25 \times 10 = $ __?__

11. $17 \times 10 = $ __?__

12. $82 \times 10 = $ __?__

13. $10 \times 54 = $ __?__

14. $10 \times 88 = $ __?__

15. $10 \times 0 = $ __?__

16. $10 \times 23 = $ __?__

17. $53 \times 10 = $ __?__

18. $44 \times 10 = $ __?__

19. $10 \times 12 = $ __?__

20. $10 \times 72 = $ __?__

21. $10 \times 30 = $ __?__

22. $77 \times 10 = $ __?__

Find more practice at **TheTimesMachine.com**. And now try the division problems on the next page. I'll do the first one for you!

Keep going! ⟶

1. 8,050 ÷ 10 = __?__

Let's Play! Since 8,050 *ends* in a zero, we know that we can just take off a zero at the end! So 8,050 ÷ 10 = **805**. Notice that we can't touch that zero in the middle of the number. Done!

Answer: 8,050 ÷ 10 = 805

2. 80 ÷ 10 = __?__

3. 50 ÷ 10 = __?__

4. 100 ÷ 10 = __?__

5. 120 ÷ 10 = __?__

6. 110 ÷ 10 = __?__

7. 630 ÷ 10 = __?__

8. 900 ÷ 10 = __?__

9. 840 ÷ 10 = __?__

10. 1,080 ÷ 10 = __?__

11. 290 ÷ 10 = __?__

12. 550 ÷ 10 = __?__

13. 1,180 ÷ 10 = __?__

14. 170 ÷ 10 = __?__

15. 10 ÷ 10 = __?__

16. 370 ÷ 10 = __?__

17. 90 ÷ 10 = __?__

18. 1,720 ÷ 10 = __?__

19. 780 ÷ 10 = __?__

20. 990 ÷ 10 = __?__

21. 0 ÷ 10 = __?__

22. 280 ÷ 10 = __?__

You probably don't need to spend much time on the 10's, so feel free to move on to the 11's. Great job!

(Answers on page 222.)

THE ELEVENS

Most of the 11's are really easy—see the pattern? When we multiply 11 by a single-digit number, we can just take that number and put it in both the tens *and* the ones digits' spots. For example:

$11 \times 2 = 22$

tens place   ones place

$11 \times 8 = 88$

tens place   ones place

Not so bad, right? Let's do this!

11 × 1 = 11
(See p. 73 for this one!)

11 × 2 = 22
(See p. 76 for this one!)

11 × 3 = 33

Fact Family

11 × 3 = 33

3 × 11 = 33

33 ÷ 11 = 3

33 ÷ 3 = 11

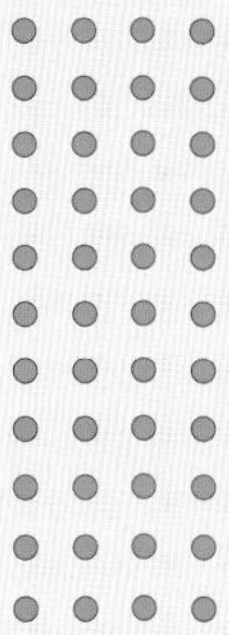

11 × 4 = 44

Fact Family

11 × 4 = 44

4 × 11 = 44

44 ÷ 11 = 4

44 ÷ 4 = 11

$$11 \times 5 = 55$$

Fact Family

$11 \times 5 = 55$

$5 \times 11 = 55$

$55 \div 11 = 5$

$55 \div 5 = 11$

$$11 \times 6 = 66$$

Fact Family

$11 \times 6 = 66$

$6 \times 11 = 66$

$66 \div 11 = 6$

$66 \div 6 = 11$

$$11 \times 7 = 77$$

Fact Family

$11 \times 7 = 77$

$7 \times 11 = 77$

$77 \div 11 = 7$

$77 \div 7 = 11$

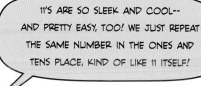

11'S ARE SO SLEEK AND COOL-- AND PRETTY EASY, TOO! WE JUST REPEAT THE SAME NUMBER IN THE ONES AND TENS PLACE, KIND OF LIKE 11 ITSELF!

$11 \times 8 = 88$

Fact Family

$11 \times 8 = 88$

$8 \times 11 = 88$

$88 \div 11 = 8$

$88 \div 8 = 11$

$11 \times 9 = 99$

Fact Family

$11 \times 9 = 99$

$9 \times 11 = 99$

$99 \div 11 = 9$

$99 \div 9 = 11$

BECAUSE 9 IS THE LARGEST SINGLE-DIGIT NUMBER, THIS IS THE LAST MULTIPLE OF 11 THAT WILL BE A *REPEAT* OF THE SAME DIGIT IN THE ONES AND TENS PLACE.

WHAT? WHAT'S GOING TO HAPPEN NOW?

DON'T WORRY--AND GUESS WHAT? TEN IS NEXT, YOUR FAVORITE!

$11 \times 10 = 110$

Fact Family

$11 \times 10 = 110$

$10 \times 11 = 110$

$110 \div 11 = 10$

$110 \div 10 = 11$

$$11 \times 11 = 121$$

Fact Family

11 × 11 = 121

121 ÷ 11 = 11

(It's a square number!)

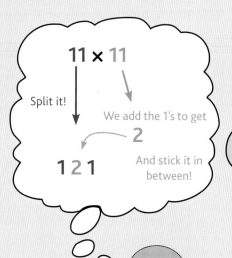

11 × 11

Split it!

We add the 1's to get

2

1 2 1

And stick it in between!

OOH, THIS IS LIKE IF ONE OF THE 11'S GOT ITS 1'S ADDED TOGETHER TO MAKE 2, AND THEN WE STUCK IT IN THE MIDDLE OF THE OTHER 11!

AS LONG AS YOU KNOW THAT'S NOT WHAT THE NUMBERS ARE ACTUALLY DOING, IT'S A FUN THING TO IMAGINE! ALSO, YOU COULD REMEMBER THAT 11 X 11 MEANS WE WANT *ONE MORE 11* THAN 11 X 10, SO WE COULD ADD 11 TO 110 AND GET 121.

YEAH . . . I LIKE MY WAY, THOUGH.

$$11 \times 12 = 132$$

(See p. 133!)

ALL THESE 11'S ARE MAKING ME HUNGRY . . . FOR FRENCH FRIES. . . .

HOW ABOUT CARROT STICKS INSTEAD?

GAME TIME!

Time to practice some multiplication facts: 11 × 0 to 11 × 11, the ones we've learned so far. Remember, to multiply any number 1–9 times 11, we just write the digit twice. And you'll be practicing 11 × 11 a few times. I'll do the first one for you!

1. 11 × 7 = _?_

Let's Play: Since 7 is a single digit, we just repeat the 7 in the ones and tens place: 77. Done!

Answer: 11 x 7 = 77

2. 3 × 11 = _?_   3. 1 × 11 = _?_   4. 6 × 11 = _?_   5. 11 × 11 = _?_

6. 11 × 2 = _?_   7. 11 × 5 = _?_   8. 11 × 11 = _?_   9. 8 × 11 = _?_

10. 4 × 11 = _?_   11. 11 × 11 = _?_   12. 7 × 11 = _?_   13. 10 × 11 = _?_

14. 11 × 11 = _?_   15. 11 × 0 = _?_   16. 11 × 8 = _?_   17. 11 × 9 = _?_

Find more practice at **TheTimesMachine.com**. And now try the division problems on the next page. I'll do the first one for you!

Keep going! ⟶

            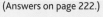

1. 121 ÷ 11 = __?__

Let's Play: Time to stand on our heads and rethink this division problem as multiplication with something missing! So it becomes 11 x ? = 121. Hey, in #5 on the previous page, we did 11 x 11 = 121. And that means the missing number is 11. Done!

Answer: 121 ÷ 11 = 11

2. 44 ÷ 4 = __?__

3. 99 ÷ 9 = __?__

4. 88 ÷ 11 = __?__

5. 121 ÷ 11 = __?__

6. 11 ÷ 1 = __?__

7. 77 ÷ 11 = __?__

8. 121 ÷ 11 = __?__

9. 110 ÷ 10 = __?__

10. 11 ÷ 11 = __?__

11. 121 ÷ 11 = __?__

12. 66 ÷ 11 = __?__

13. 0 ÷ 11 = __?__

14. 55 ÷ 11 = __?__

15. 121 ÷ 11 = __?__

16. 99 ÷ 9 = __?__

17. 121 ÷ 11 = __?__

Great job! You might not need a full week to memorize these because most of them are so easy! But you will want to memorize 11 x 11 = 121 and 11 x 12 = 132 (which you'll learn on p. 133). You can go to **TheTimesMachine.com** to download and print multiplication facts to tape on your bathroom mirror, so twice a day you'll see them when you're brushing your teeth! Then, once you feel like you know them really well, move on to the 12's. I'm so proud of you!

Have a great day!

Hi, cutie!

$$\begin{array}{r} 11 \\ \times\ 11 \\ \hline 121 \end{array}$$

$$\begin{array}{r} 11 \\ \times\ 12 \\ \hline 132 \end{array}$$

(Answers on page 222.)

×

THE TWELVES

's

Remember how we broke numbers with hammers and split the bagels in Chapter 3, using the Distributive Property to rewrite multiplication problems as two easier ones? That's the trick we're going to use for the 12's, because after all, 12 = 10 + 2, and the 10's are really easy, and so are the 2's!

Let's do an example; we'll multiply 12 × 7. If we use the "Hammer" Trick from p. 44 to break up 12 into 10 + 2, we just have to do these easier problems: 10 × 7 = **70** and 2 × 7 = **14**, and then add 70 + 14 to get our answer, **84**. It's a great way to exercise mental math skills—you know, doing math in your head!

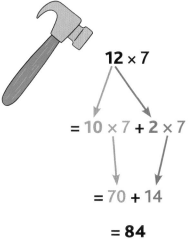

12 × 7

= 10 × 7 **+ 2** × 7

= 70 **+ 14**

= **84**

I'll show you what this looks like for each of the 12's facts—but if you did the exercises on p. 45, I bet you can figure out the facts right now on your own! I'm so proud of how far you've come.

12 × 1 = 12	12 × 2 = 24
(See p. 73 for this one!)	(See p. 76 for this one!)

The Core of the Times Machine  **129**

12 × 3 = 36

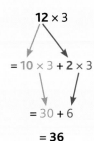

Let's break the 12 with a hammer!

12 × 3

= 10 × 3 + **2** × 3

= 30 + 6

= **36**

Fact Family

12 × 3 = 36

3 × 12 = 36

36 ÷ 12 = 3

36 ÷ 3 = 12

---

12 × 4 = 48

Let's break the 12 with a hammer!

12 × 4

= 10 × 4 + **2** × 4

= 40 + 8

= **48**

Fact Family

12 × 4 = 48

4 × 12 = 48

48 ÷ 12 = 4

48 ÷ 4 = 12

---

12 × 5 = 60

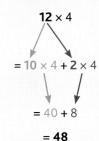

IT'S COOL HOW THE 12'S MULTIPLES JUMP RIGHT OVER THE 50'S! THEY GO STRAIGHT FROM 48 TO 60!

Let's break the 12 with a hammer!

12 ×5

= 10 × 5 + **2** × 5

= 50 + 10

= **60**

Fact Family

12 × 5 = 60

5 × 12 = 60

60 ÷ 12 = 5

60 ÷ 5 = 12

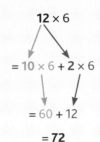

12 × 6 = 72

Let's break the 12 with a hammer!

12 × 6

= 10 × 6 + **2** × 6

= 60 + 12

= **72**

Fact Family

12 × 6 = 72

6 × 12 = 72

72 ÷ 12 = 6

72 ÷ 6 = 12

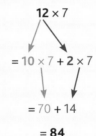

12 × 7 = 84

Let's break the 12 with a hammer!

12 × 7

= 10 × 7 + **2** × 7

= 70 + 14

= **84**

Fact Family

12 × 7 = 84

7 × 12 = 84

84 ÷ 12 = 7

84 ÷ 7 = 12

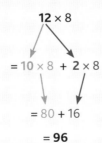

12 × 8 = 96

Let's break the 12 with a hammer!

12 × 8

= 10 × 8 + **2** × 8

= 80 + 16

= **96**

Fact Family

12 × 8 = 96

8 × 12 = 96

96 ÷ 12 = 8

96 ÷ 8 = 12

12 × 9 = 108

Let's break the 12 with a hammer!

12×9

$= 10 \times 9 + \mathbf{2} \times 9$

$= 90 + 18$

$= \mathbf{108}$

Fact Family
12 × 9 = 108
9 × 12 = 108
108 ÷ 12 = 9
108 ÷ 9 = 12

WAIT, WHAT ABOUT THE 9'S TRICK?

WELL, WE HAVE 10 FINGERS, SO THAT TRICK ONLY WORKS WHEN WE MULTIPLY 9 BY A SINGLE-DIGIT NUMBER, LIKE 4, 5, OR EVEN 9. BUT YOU KNOW WHAT IS COOL? THE DIGITS OF THE ANSWER STILL ADD UP TO 9! CHECK IT OUT: 9 X 12 = 108, AND IF WE ADD UP THE DIGITS: 1 + 0 + 8 = 9. YEP, THEY STILL ADD UP TO BE 9! REMEMBER--THAT "ADDING UP THE DIGITS" TRICK ONLY WORKS FOR 9.

WHAT ABOUT 9 X 11? THAT ANSWER IS 99, AND 9 + 9 = *18*, NOT 9.

GOOD POINT! BUT NOTICE THAT FOR *18*, WE CAN ADD *THOSE* DIGITS AND WE GET 1 + 8 = 9. FOR MANY OF THE BIGGER NUMBERS, YOU HAVE TO ADD THE DIGITS OF *THAT* ANSWER IN ORDER TO GET THE 9 YOU'RE LOOKING FOR.

MIND = BLOWN.

12 × 10 = 120

(See p. 119 for this one!)

12 × 11 = 132

Let's break the 12 with a hammer!

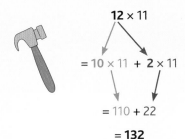

12 × 11

= **10** × 11 + **2** × 11

= 110 + 22

= **132**

Fact Family

12 × 11 = 132

11 × 12 = 132

132 ÷ 12 = 11

132 ÷ 11 = 12

---

12 × 12 = 144

We can still break a 12 with a hammer!

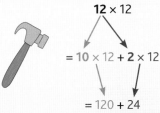

12 × 12

= **10** × 12 + **2** × 12

= 120 + 24

= **144**

Fact Family

12 × 12 = 144

144 ÷ 12 = 12

(It's a square number!)

OOH, THIS IS THE [VE]RY LAST ONE! AND IT'S A SQUARE NUMBER-- COOL!

BUT WAIT, THERE ARE TWO 12'S--I'M WORRIED THE HAMMER TRICK WILL CONFUSE ME!

HERE'S A GOOD WAY TO THINK ABOUT IT: WE ALREADY KNOW THAT 12 X 10 = 120, RIGHT? THEN WE JUST NEED *TWO MORE 12'S* TO GET 12 X 12. AND HOW MUCH IS TWO MORE 12'S? 24! AND 120 + 24 = **144**, SO THAT'S OUR ANSWER: 12 X 12 = 144.

CONGRATULATIONS!!! We've made it through all of our facts!

GAME TIME!

It's time to practice the 12's multiplication facts! Remember, to multiply any number by 12, we can just multiply it by 10, and also by 2, and then add them together. Eventually it's best to memorize them, which will happen the more you practice, if you're paying attention. I'll do the first one for you!

1. $12 \times 9 = \underline{\ ?\ }$

Let's Play: We'll use a hammer (the Distributive Property) to split the 12 into 10 + 2. First, we'll multiply 10 x 9 = **90,** and then we'll multiply 2 x 9 = **18,** and then we add them together: **90 + 18** = 108. Done!

Answer: 12 x 9 = 108

2. $3 \times 12 = \underline{\ ?\ }$

3. $1 \times 12 = \underline{\ ?\ }$

4. $6 \times 12 = \underline{\ ?\ }$

5. $12 \times 11 = \underline{\ ?\ }$

6. $12 \times 2 = \underline{\ ?\ }$

7. $12 \times 5 = \underline{\ ?\ }$

8. $12 \times 12 = \underline{\ ?\ }$

9. $8 \times 12 = \underline{\ ?\ }$

10. $2 \times 12 = \underline{\ ?\ }$

11. $9 \times 12 = \underline{\ ?\ }$

12. $12 \times 7 = \underline{\ ?\ }$

13. $10 \times 12 = \underline{\ ?\ }$

14. $12 \times 8 = \underline{\ ?\ }$

15. $12 \times 0 = \underline{\ ?\ }$

16. $11 \times 12 = \underline{\ ?\ }$

17. $5 \times 12 = \underline{\ ?\ }$

18. $4 \times 12 = \underline{\ ?\ }$

19. $9 \times 12 = \underline{\ ?\ }$

20. $7 \times 12 = \underline{\ ?\ }$

21. $12 \times 3 = \underline{\ ?\ }$

I recommend doing those a few times! Also find more practice at **TheTimesMachine.com.** Then, <u>once you're pretty comfortable with them,</u> try the division problems on the next page. I'll do the first one for you!

(Answers on page 222.)

1. $84 \div 12 =$ __?__

Let's Play: Time to stand on our heads and rethink this division problem as multiplication with something missing! So it becomes 12 x ? = 84. Hey, in #12 on the previous page, we did 12 x 7 = 84. And that means the missing number is 7. Done!

Answer: $84 \div 12 = 7$

2. $48 \div 4 =$ __?__   3. $108 \div 9 =$ __?__   4. $48 \div 12 =$ __?__   5. $96 \div 8 =$ __?__

6. $12 \div 1 =$ __?__   7. $84 \div 7 =$ __?__   8. $120 \div 10 =$ __?__   9. $24 \div 2 =$ __?__

10. $72 \div 12 =$ __?__   11. $24 \div 12 =$ __?__   12. $12 \div 12 =$ __?__   13. $132 \div 11 =$ __?__

14. $84 \div 12 =$ __?__   15. $60 \div 5 =$ __?__   16. $132 \div 12 =$ __?__   17. $60 \div 12 =$ __?__

18. $0 \div 12 =$ __?__   19. $96 \div 12 =$ __?__   20. $72 \div 6 =$ __?__   21. $36 \div 12 =$ __?__

Great job! Read the 12's section every day for a week, and practice these problems every day for a week, too. You can also go to **TheTimesMachine.com** to download and print multiplication facts to tape on your bathroom mirror, so twice a day you'll see them when you're brushing your teeth! Then, once you feel like you know them really well, try using flash cards to see which multiplication facts you still need to practice, and just tape THOSE facts to your bathroom mirror for a while, until you know them all cold. You might want to do two rounds of the 12's—first learn 12 × 2 to 12 × 6, and then the next week do 12 × 7 to 12 × 12. You got this—and I'm so proud of you!

(Answers on page 222.)

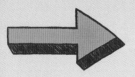 # GAME TIME!

Now that you've spent several weeks practicing your multiplication and division facts, see how many of these you can do in ten minutes!
Try printing a copy at TheTimesMachine.com
(without the question marks) that you can write on, too!

1. 3 × 12 = ?
2. 80 ÷ 8 = ?
3. 6 × 5 = ?
4. 24 ÷ 6 = ?
5. 7 × 2 = ?

6. 66 ÷ 11 = ?
7. 7 × 9 = ?
8. 81 ÷ 9 = ?
9. 6 × 7 = ?
10. 48 ÷ 4 = ?

11. 8 × 6 = ?
12. 60 ÷ 12 = ?
13. 9 × 4 = ?
14. 0 ÷ 7 = ?
15. 5 × 12 = ?

16. 54 ÷ 9 = ?
17. 8 × 4 = ?
18. 24 ÷ 12 = ?
19. 4 × 6 = ?
20. 36 ÷ 12 = ?

21. 8 × 0 = ?
22. 70 ÷ 7 = ?
23. 6 × 3 = ?
24. 18 ÷ 2 = ?
25. 12 × 10 = ?

26. 64 ÷ 8 = ?
27. 2 × 6 = ?
28. 32 ÷ 8 = ?
29. 7 × 8 = ?
30. 60 ÷ 6 = ?

31. 3 × 2 = ?
32. 18 ÷ 3 = ?
33. 7 × 5 = ?
34. 10 ÷ 5 = ?
35. 6 × 12 = ?

36. 36 ÷ 9 = ?
37. 4 × 7 = ?
38. 12 ÷ 4 = ?
39. 3 × 8 = ?
40. 88 ÷ 11 = ?

41. 7 × 3 = ?
42. 36 ÷ 9 = ?
43. 9 × 9 = ?
44. 24 ÷ 8 = ?
45. 5 × 1 = ?

46. 132 ÷ 12 = ?
47. 12 × 9 = ?
48. 35 ÷ 5 = ?
49. 2 × 2 = ?
50. 90 ÷ 10 = ?

51. 6 × 11 = ?
52. 28 ÷ 7 = ?
53. 3 × 3 = ?
54. 27 ÷ 3 = ?
55. 5 × 8 = ?

56. 45 ÷ 5 = ?
57. 8 × 8 = ?
58. 63 ÷ 7 = ?
59. 2 × 5 = ?
60. 96 ÷ 8 = ?

61. 1 × 0 = ?
62. 25 ÷ 5 = ?
63. 12 × 7 = ?
64. 30 ÷ 10 = ?
65. 3 × 5 = ?

66. 20 ÷ 5 = ?
67. 2 × 8 = ?
68. 144 ÷ 12 = ?
69. 7 × 7 = ?
70. 15 ÷ 5 = ?

71. 4 × 12 = ?
72. 50 ÷ 10 = ?
73. 10 × 11 = ?
74. 16 ÷ 8 = ?
75. 4 × 3 = ?

76. 21 ÷ 3 = ?
77. 5 × 5 = ?
78. 48 ÷ 8 = ?
79. 12 × 8 = ?
80. 44 ÷ 4 = ?

81. $7 \times 1 = \underline{?}$ 82. $12 \div 2 = \underline{?}$ 83. $7 \times 11 = \underline{?}$ 84. $48 \div 12 = \underline{?}$ 85. $10 \times 9 = \underline{?}$

86. $8 \div 2 = \underline{?}$ 87. $9 \times 11 = \underline{?}$ 88. $108 \div 9 = \underline{?}$ 89. $10 \times 3 = \underline{?}$ 90. $84 \div 12 = \underline{?}$

91. $5 \times 5 = \underline{?}$ 92. $14 \div 7 = \underline{?}$ 93. $6 \times 1 = \underline{?}$ 94. $30 \div 5 = \underline{?}$ 95. $12 \times 11 = \underline{?}$

96. $42 \div 6 = \underline{?}$ 97. $1 \times 1 = \underline{?}$ 98. $56 \div 7 = \underline{?}$ 99. $9 \times 10 = \underline{?}$ 100. $50 \div 5 = \underline{?}$

101. $5 \times 4 = \underline{?}$ 102. $6 \div 1 = \underline{?}$ 103. $9 \times 3 = \underline{?}$ 104. $20 \div 10 = \underline{?}$ 105. $8 \times 9 = \underline{?}$

106. $3\overline{)36}$ = ?
107. $\begin{array}{r} 5 \\ \times 2 \\ \hline ? \end{array}$
108. $6\overline{)72}$ = ?
109. $\begin{array}{r} 3 \\ \times 6 \\ \hline ? \end{array}$
110. $6\overline{)54}$ = ?

111. $\begin{array}{r} 12 \\ \times 3 \\ \hline ? \end{array}$
112. $9\overline{)18}$ = ?
113. $\begin{array}{r} 12 \\ \times 12 \\ \hline ? \end{array}$
114. $10\overline{)60}$ = ?
115. $\begin{array}{r} 10 \\ \times 1 \\ \hline ? \end{array}$

116. $11\overline{)99}$ = ?
117. $\begin{array}{r} 7 \\ \times 6 \\ \hline ? \end{array}$
118. $6\overline{)18}$ = ?
119. $\begin{array}{r} 12 \\ \times 5 \\ \hline ? \end{array}$
120. $9\overline{)63}$ = ?

121. $\begin{array}{r} 6 \\ \times 4 \\ \hline ? \end{array}$
122. $7\overline{)77}$ = ?
123. $\begin{array}{r} 4 \\ \times 4 \\ \hline ? \end{array}$
124. $11\overline{)0}$ = ?
125. $\begin{array}{r} 12 \\ \times 1 \\ \hline ? \end{array}$

126. $5\overline{)60}$ = ?
127. $\begin{array}{r} 5 \\ \times 7 \\ \hline ? \end{array}$
128. $2\overline{)10}$ = ?
129. $\begin{array}{r} 10 \\ \times 12 \\ \hline ? \end{array}$
130. $9\overline{)27}$ = ?

131. $\begin{array}{r} 6 \\ \times 6 \\ \hline ? \end{array}$
132. $8\overline{)88}$ = ?
133. $\begin{array}{r} 4 \\ \times 8 \\ \hline ? \end{array}$
134. $10\overline{)70}$ = ?
135. $\begin{array}{r} 5 \\ \times 6 \\ \hline ? \end{array}$

136. $6\overline{)12}$ = ?
137. $\begin{array}{r} 7 \\ \times 4 \\ \hline ? \end{array}$
138. $3\overline{)12}$ = ?
139. $\begin{array}{r} 12 \\ \times 6 \\ \hline ? \end{array}$
140. $4\overline{)32}$ = ?

141. $\begin{array}{r} 3 \\ \times 4 \\ \hline ? \end{array}$
142. $5\overline{)55}$ = ?
143. $\begin{array}{r} 6 \\ \times 8 \\ \hline ? \end{array}$
144. $4\overline{)24}$ = ?
145. $\begin{array}{r} 8 \\ \times 7 \\ \hline ? \end{array}$

146. $2\overline{)20}$ = ?
147. $\begin{array}{r} 6 \\ \times 2 \\ \hline ? \end{array}$
148. $12\overline{)84}$ = ?
149. $\begin{array}{r} 3 \\ \times 1 \\ \hline ? \end{array}$
150. $8\overline{)56}$ = ?

151. $\begin{array}{r} 2 \\ \times 3 \\ \hline ? \end{array}$
152. $7\overline{)21}$ = ?
153. $\begin{array}{r} 9 \\ \times 12 \\ \hline ? \end{array}$
154. $3\overline{)30}$ = ?
155. $\begin{array}{r} 0 \\ \times 1 \\ \hline ? \end{array}$

Keep going! $\longrightarrow$

156. $3\overline{)33}$?

157. $\begin{array}{r} 8 \\ \times\ 3 \\ \hline ? \end{array}$

158. $9\overline{)45}$?

159. $\begin{array}{r} 11 \\ \times\ 11 \\ \hline ? \end{array}$

160. $7\overline{)49}$?

161. $\begin{array}{r} 9 \\ \times\ 5 \\ \hline ? \end{array}$

162. $7\overline{)35}$?

163. $\begin{array}{r} 3 \\ \times\ 7 \\ \hline ? \end{array}$

164. $6\overline{)48}$?

165. $\begin{array}{r} 4 \\ \times\ 9 \\ \hline ? \end{array}$

166. $12\overline{)108}$?

167. $\begin{array}{r} 7 \\ \times\ 12 \\ \hline ? \end{array}$

168. $3\overline{)24}$?

169. $\begin{array}{r} 4 \\ \times\ 5 \\ \hline ? \end{array}$

170. $12\overline{)72}$?

171. $\begin{array}{r} 9 \\ \times\ 6 \\ \hline ? \end{array}$

172. $3\overline{)15}$?

173. $\begin{array}{r} 8 \\ \times\ 2 \\ \hline ? \end{array}$

174. $2\overline{)24}$?

175. $\begin{array}{r} 2 \\ \times\ 4 \\ \hline ? \end{array}$

176. $3\overline{)6}$?

177. $\begin{array}{r} 12 \\ \times\ 2 \\ \hline ? \end{array}$

178. $4\overline{)28}$?

179. $\begin{array}{r} 5 \\ \times\ 9 \\ \hline ? \end{array}$

180. $10\overline{)110}$?

181. $\begin{array}{r} 2 \\ \times\ 7 \\ \hline ? \end{array}$

182. $3\overline{)9}$?

183. $11\overline{)132}$?

184. $6\overline{)30}$?

185. $\begin{array}{r} 8 \\ \times\ 12 \\ \hline ? \end{array}$

186. $2\overline{)16}$?

187. $\begin{array}{r} 12 \\ \times\ 4 \\ \hline ? \end{array}$

188. $7\overline{)42}$?

189. $\begin{array}{r} 10 \\ \times\ 0 \\ \hline ? \end{array}$

190. $2\overline{)14}$?

191. $\begin{array}{r} 3 \\ \times\ 9 \\ \hline ? \end{array}$

192. $4\overline{)40}$?

193. $\begin{array}{r} 8 \\ \times\ 5 \\ \hline ? \end{array}$

194. $4\overline{)20}$?

195. $\begin{array}{r} 9 \\ \times\ 7 \\ \hline ? \end{array}$

196. $11\overline{)33}$?

197. $\begin{array}{r} 2 \\ \times\ 12 \\ \hline ? \end{array}$

198. $2\overline{)22}$?

199. $\begin{array}{r} 11 \\ \times\ 12 \\ \hline ? \end{array}$

200. $4\overline{)8}$?

201. $\begin{array}{r} 5 \\ \times\ 3 \\ \hline ? \end{array}$

202. $4\overline{)36}$?

203. $\begin{array}{r} 11 \\ \times\ 8 \\ \hline ? \end{array}$

204. $12\overline{)96}$?

205. $\begin{array}{r} 6 \\ \times\ 9 \\ \hline ? \end{array}$

206. $2\overline{)6}$?

207. $\begin{array}{r} 9 \\ \times\ 8 \\ \hline ? \end{array}$

208. $12\overline{)120}$?

209. $\begin{array}{r} 10 \\ \times\ 10 \\ \hline ? \end{array}$

210. $11\overline{)121}$?

GREAT JOB GETTING THROUGH THE CORE OF THE TIMES MACHINE!

As you keep practicing these and solidify them in your memory, you're becoming more and more powerful—in math and in life! Now let's see what we can *do* with all this new power. . . .

(Answers at **TheTimesMachine.com**.)

Chapter 6

The Multiplication Boogie:
Order of Operations, Number Properties, and More!

Hungry Pandas: The Order of Operations

Now that we know our facts, we might be asked to solve something like $9 + 3 \times 2 = ?$ It's tempting to add the $9 + 3$ first, but that would give the wrong answer. Crazy, right? How about something like $6 \times (3 \times 10) = ?$ or $80 \times 5,000 = ?$ Soon we'll see how simple it can be to get the right answers, and some pandas and a little dancing are going to help us.

I LIKE THE TITLE OF THIS CHAPTER! "THE MULTIPLICATION BOOGIE."

I MEAN, BOOGERS ARE AWESOME, BUT--

NOT BOOGERS-- "BOOGIE."

IN MY WORLD, THOSE ARE THE SAME THING.

IT MEANS "DANCE."

The **Order of Operations,** sometimes called **PEMDAS,** is the order we must use when we solve math expressions so we get the right answer:

1. **P**arentheses
2. **M**ultiplication & **D**ivision (whichever comes first, left to right)
3. **A**ddition & **S**ubtraction (whichever comes first, left to right)

PEMDAS: **P**andas **E**at **M**ustard on **D**umplings and **A**pples with **S**pice!

In any math expression, first we solve anything that might be inside **P**arentheses, then we look for any **M**ultiplication & **D**ivision and do those, left to right, and THEN we do the same for any **A**ddition & **S**ubtraction, left to right. Not so bad, huh?

For example, to solve $4 + 6 \div 2 = ?$, there are no parentheses, but we should do the division first, *then* the addition, and we get the right answer, 7. If we'd added $4 + 6$ first, we'd get the wrong answer, 5. We'll do more on the next page!

Our pandas help us remember the correct Order of Operations: **P**andas **E**at **M**ustard on **D**umplings and **A**pples with **S**pice (PEMDAS). Notice that the mustard & dumplings **go together,** and the apples & spice **go together**—like how the multiplication & division are in the same step, and addition & subtraction are in the same step.

QUICK NOTE

When you get older, you'll learn the full "PEMDAS" rule, which also includes something called *exponents.* So the "E" in "eat" stands for **E**xponents! If you're curious, you can see more about that on p. 21 of my book *Kiss My Math.*

It's true that the word "multiplication" comes before "division" in our PEMDAS sentence, but don't be fooled into thinking that multiplication always comes before division: That's why the mustard is spread on the dumplings—they go *together*! If we see multiplication and division, we do whichever comes *first*:

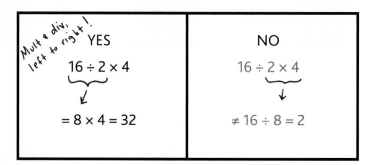

Mult & div, left to right!

YES	NO
$16 \div 2 \times 4$	$16 \div 2 \times 4$
$= 8 \times 4 = 32$	$\neq 16 \div 8 = 2$

If somebody wanted us to do the multiplication first here, they'd have to put parentheses around it, like this: $16 \div (2 \times 4)$. See what I mean?

Here's another example: If we're told to solve $4 + 3 \times 7 = ?$, what do we do first? Well, thinking about our pandas—there are no Parentheses, just Addition and Multiplication, right? We know the pandas definitely eat the mustard (Multiplication) *before* the apples (Addition) they have for dessert, so we have to multiply 3×7 first, which becomes 21. And we're left with addition, $4 + 21$, which is easy: **25**!

$$4 + 3 \times 7$$
$$= 4 + 21$$
$$= 25$$

But what if that same problem had parentheses, like this: $(4 + 3) \times 7 = ?$ First we'd do what's in the **P**arentheses, right? We'd add $4 + 3$ so it becomes 7, we'd be left with 7×7, and *then* we'd **M**ultiply 7×7 to get **49**.

$$(4 + 3) \times 7$$
$$= 7 \times 7$$
$$= \mathbf{49}$$

Wow, 25 and 49 are pretty different answers—and the only thing that changed was the *parentheses*!

Dropping Parentheses

When we put parentheses around numbers, it means, "Do this first!" And parentheses do a great job of keeping things separate. But once we solve what's inside the parentheses, we can drop them. I mean, in the example above, we could have written:

But we don't need the parentheses anymore!

$$(4 + 3) \times 7$$
$$= (7) \times 7$$
$$= 49$$

Once whatever was inside is just a **single number,** we usually drop the parentheses because we don't need them anymore.

**Solve these expressions, using our PEMDAS "Panda" rule from p. 141.
I'll do the first one for you!**

1. $10 \div 2 \times (9 - 5 + 4) = ?$

Let's Play: Whoa! This might look a little scary at first, but we'll let our pandas help us. Right away, we see parentheses, so let's solve the *inside* of those: $9 - 5 + 4 = 8$. Great! Now our expression looks like this: $10 \div 2 \times 8 = ?$ See? We stuck in "8" for where the parentheses stuff used to be. (See the Quick Note on p. 143 for why we can drop the parentheses.) Looking better already! Next, we are left with multiplication and division, which we should do left to right: Since $10 \div 2 = 5$, the whole thing is now just 5×8, which we know from our multiplication facts: 40. Phew, done!

Answer: 40

2. $(9 + 3) \times 2 = ?$

3. $9 + 3 \times 2 = ?$

4. $30 \div (3 \times 5) = ?$

5. $30 \div 3 \times 5 = ?$

6. $10 - 2 + 6 - 1 = ?$

7. $10 - (2 + 6) - 1 = ?$

8. $10 - 2 + (6 - 1) = ?$

9. $9 \times 0 + 8 = ?$

10. $9 \times (0 + 8) = ?$

11. $6 \times (4 \div 2) \times 3 = ?$

12. $6 \times 4 \div (2 \times 3) = ?$

13. $6 \times 4 \div 2 \times 3 = ?$

(Answers on page 222.)

The Multiplication Boogie:
The Commutative and Associative Properties

When we have <u>multiplication only</u>, the numbers can . . .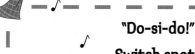

"Do-si-do!" **Switch spots**	**"Change partners!"** **Move the parentheses**

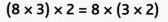

"Do-si-do!"

$8 \times 9 = 9 \times 8$

Both of these have the

same answer: **72**!

This is easy to remember because our

fact families remind us this is true.

This is an example of the

Commutative Property.

(See p. 147 for more.)

"Change partners!"

$(8 \times 3) \times 2 = 8 \times (3 \times 2)$

We won't move the numbers—just the

parentheses to show a different order.

So we can do the 8×3 part first:

$(8 \times 3) \times 2$

$= 24 \times 2 = \mathbf{48}$

. . . or the 3×2 part first:

$8 \times (3 \times 2)$

$= 8 \times 6 = \mathbf{48}$

. . . and we get the same answer!

In other words:

$(8 \times 3) \times 2 = 8 \times (3 \times 2)$

This is an example of the

Associative Property.

(See p. 147 for more.)

THOSE CUTE PARENTHESES MAKE THE NUMBERS LOOK LIKE DANCE PARTNERS ALL CUDDLED UP AND DANCING CLOSE, DON'T YOU THINK? IT'S SO SWEET.

SURE, UNTIL THEY MOVE PARENTHESES AND CHANGE PARTNERS. HAVEN'T YOU BEEN LISTENING?

NOW LET'S SEE THE DEFINITIONS OF THESE TWO DANCE MOVES.

The **Commutative Property of Multiplication** says if the *only* operation between numbers is multiplication, then we can change the order of the factors and the product stays the same. For example:

$$5 \times 8 = 8 \times 5$$

They both equal 40!

THAT'S LIKE THE FACT FAMILIES. LIKE HOW 2 X 8 = 8 X 2. THEY'RE BOTH 16, DUH.

YEAH! AND IF WE'RE DANCING, THAT'S LIKE YOU AND ME SWITCHING SPOTS--A LITTLE DO-SI-DO!

??? WHAT'S IT CALLED? ???

The **Associative Property of Multiplication** says if the *only* operation between numbers is multiplication, then we can move the parentheses around and we'll still get the same answer. In other words, the grouping doesn't matter. For example:

$$5 \times (2 \times 7) = (5 \times 2) \times 7$$

They both equal 70!

CHANGE PARTNERS!

Notice that the above pink expression $(5 \times 2) \times 7$ is a lot easier to solve, because doing the parentheses first, we get 10, and then 10×7 is easy! In the blue expression, $5 \times (2 \times 7)$, first we'd multiply $2 \times 7 = 14$, and then figure out that $5 \times 14 = 70$. Not bad, but not as easy as the pink one, right? That's one example of why we might *want* to let the numbers do a little dance and change partners using the Associative Property.

Remember, the Commutative and Associative "dance moves" won't work if we also have addition, subtraction, or division involved—it has to be ONLY multiplication.

YES	NO
$8 \times 4 = 4 \times 8$	$8 \div 4 \neq 4 \div 8$
Commutative Property	$(6 + 3) \times 2 \neq 6 + (3 \times 2)$
$(6 \times 3) \times 2 = 6 \times (3 \times 2)$	
Associative Property	

There is also the *Commutative Property of Addition* and the *Associative Property of Addition*. So this is true:

$$3 + 4 = 4 + 3 \qquad (9 + 5) + 2 = 9 + (5 + 2)$$

But as soon as we start *mixing* addition and multiplication together, we cannot move the parentheses around. Just remember: When it's <u>only multiplication, or only addition</u>, the numbers can dance all around each other and we'll always get the right answer.

GAME TIME!

Use the Associative Property to turn the following problems into one of the 0–12 multiplication facts we've learned, and then solve it! I'll do the first one for you!

1. (12 × 3) × 4 = ?

Let's Play: Hmm, we could try to do 36 × 4, but since the only operation here is multiplication, we can move those parentheses. Let's try 12 × (3 × 4) = ? And lookie there, since 3 × 4 = 12, this problem now becomes 12 × 12, and we know that from our multiplication facts—it's 144!

Answer: 12 × 12 = 144

2. 4 × (3 × 5) = ?

3. 3 × (2 × 9) = ?

4. (6 × 6) × 2 = ?

5. (8 × 3) × 3 = ?

6. 2 × (5 × 11) = ?

7. (10 × 2) × 5 = ?

8. (7 × 2) × 4 = ?

9. 6 × (2 × 7) = ?

10. 3 × (2 × 11) = ?

CHANGE PARTNERS!

(Answers on page 222.)

A Bigger Boogie: Multiplication with Multiples of 10's, 100's, and More!

In Chapter 8, we'll be using the multiplication facts we just learned to multiply bigger numbers. For now, let's check out a fun and easy way to multiply numbers that *end in zeros*. You know how it's super easy to multiply by 10, because we just add a zero to change its *place value*? (See p. 152 for why that works!)

Now we'll do that same type of trick to easily multiply factors like 6 × 400.

STEP BY STEP:

The "Zeros Trick":
How to Multiply Times Multiples of 10, 100, 1,000, and More!

Here's a trick to multiply any whole number times multiples of 10, and also multiples of 100, 1,000, and more! It all comes down to keeping track of the correct *place value* by using the right number of zeros. I'll show you with the example **6 × 400 = ?**

Step 1. Notice how many zeros are at the ends of the factors, then drop them and multiply the digits that are left.
In this case, we notice there are <u>two</u> zeros from the 400, so we drop those.
Next, we just multiply 6 × 4 = **24**.
Step 2. We stick the same number of zeros back on.
We had a total of <u>two</u> zeros from the factors, right? So we put those back on, to keep the correct place value: 2400.
Step 3. For products that are 1,000 or bigger, we'll want to add commas.
In this case, our answer becomes: **2,400**.
Done! **6 × 400 = 2,400**

Here's another example: 80 × 5,000 = ?
Step 1. Drop the (four) zeros from the 80 and 5000, and just multiply 8 × 5 = **40**.
Step 2. Now we add those four zeros back on. (Notice we have one zero in 40, so adding four zeros gives a total of <u>five</u> zeros now!) We get: 400000.
Step 3. Add commas where needed: 400,000. (See the Quick Note on the next page!)
Done! **80 × 5,000 = 400,000**

Place Value!

When we add a zero to a whole number, what we're really doing is moving the *place value* of that whole number by one. For example, to multiply 7×10, we want to move the place value of 7 by one spot, so we add one zero: **70**. Now the 7 is in the *tens* place, see? Or if instead we multiply 7×100, we want to move 7's place value by two spots, so we add two zeros, putting the 7 in the *hundreds* place: **700**. This is what we were actually doing on p. 151.

And the same works for division! Just like we did on p. 120, the reason we take off a zero for $500 \div 10 = 50$ is to move 5's place value by one spot, from the hundreds place to the tens place. And now you know what's really going on, you rock star!

Watch Out!

Once we learn how to multiply bigger numbers like $90 \times 801 = ?$, we would only drop the zero from the 90—we would NOT drop the zero between the 8 and the 1. It's only the zeros at the *ends* of the factors that we can drop and then put back on at the end!

Commas!

Any time we write a number that is four or more digits, we should write a comma between the thousand and hundred places, and then *every three* places from there, moving to the left, if the number is big enough. For example:

3,500   67,824   3,456,231   1,000   867,342,463,674

By the way, not everybody uses commas for four-digit numbers, but we're supposed to!

GAME TIME!

**Multiply these numbers using the Step by Step shortcut from p. 151.
I'll do the first one for you!**

1. 90 × 400 = ?

Let's Play: Using the steps, we'll first notice there are *three* zeros total—one from the 90 and two from the 400, right? Then we'll ignore the zeros and multiply 9 × 4 = 36. Next, we'll stick the three zeros back on: 36000, and finally, we'll stick in our comma: 36,000. Done!

Answer 90 × 400 = 36,000

2. 20 × 7 = ___?___

3. 4 × 90 = ___?___

4. 5 × 90 = ___?___

5. 3 × 400 = ___?___

6. 20 × 80 = ___?___

7. 8 × 400 = ___?___

8. 500 × 500 = ___?___

9. 600 × 500 = ___?___

10. 900 × 7,000 = ___?___

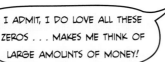

I ADMIT, I DO LOVE ALL THESE ZEROS . . . MAKES ME THINK OF LARGE AMOUNTS OF MONEY!

(Answers on page 222.)

Chapter 7

What Should We Do with Leftover Bananas?
Remainder Division

More Monkeys and Leftover Bananas—Remainders!

In Chapter 4, we talked about division as "fair sharing" since it's splitting things into equal groups, right? But things aren't always easy to split up evenly. For example, how would we divide 7 bananas among 3 monkeys?

WAIT. ARE YOU SAYING WE CAN'T GIVE THE SAME NUMBER OF BANANAS TO EACH MONKEY?

I'LL EAT THAT.

WE CAN, BUT WE'LL END UP WITH A BANANA LEFT OVER. IT'S CALLED A *REMAINDER.*

→ remainder!

$$7 \div 3 = 2 \text{ R } 1$$

The **remainder** is the amount left over, after division. For example, $7 \div 3 = \textbf{2 R1}$. That's because we can divide 7 into 3 groups of **2** each, with a remainder of **1**.

I **LOVE** REMINDERS! LIKE IF I FORGET TO BRUSH MY TEETH, MY MOM ALWAYS GIVES ME A REMINDER SO I DON'T FORGET THE NEXT TIME.

NOT REMINDERS--**REMAINDERS**. AS IN, "IT FEELS LIKE THE **MAIN** POINT OF THIS LESSON IS TO MAKE ME HUNGRY FOR BANANAS."

Here's another example: $14 \div 3 = ?$ Hmm, if we're making 3 equal groups, how many bananas could go in each group? Well, we *can't* put 5 bananas in each group, because $3 \times 5 = 15$, and we don't have 15 bananas—we only have 14. So I guess we can only put **4** bananas in each of the 3 groups, right? And that means we'll have 2 bananas left over, see?

$$14 \div 3 = ?$$

We put 4 in each of the 3 groups . . .

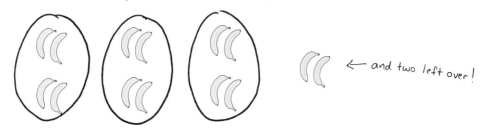

← and two left over!

With 4 bananas in each group and 2 left over, we get the answer: $14 \div 3 = 4$ R2.

Notice that when we do a division problem like 14 ÷ 3 = 4 R2, we can write the groups we make as a *picture* with the "leftovers," and also as a multiplication sentence!

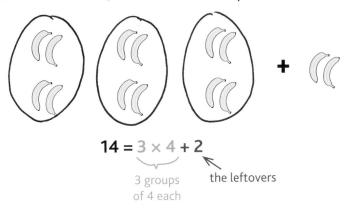

$14 = 3 \times 4 + 2$

3 groups   the leftovers
of 4 each

So the above picture can describe the multiplication sentence 14 = 3 × 4 + 2 and also the *division sentence* 14 ÷ 3 = 4 R2. Do you see how we can write these two math sentences from the same picture? Read that again until it makes sense to you!

By the way, because the Commutative Property tells us that
3 × 4 = 4 × 3, we also could have written the above banana
sentence as 14 = 4 × 3 + 2.

QUICK
NOTE

OOH, "MATH SENTENCES." I LOVE
THESE! THEY DESCRIBE A STORY
WITH NUMBERS . . . AND THIS TIME
WITH LEFTOVERS, TOO!

I PREFER THE
BANANA SENTENCES.

WELL, WE WON'T
ALWAYS USE BANANAS.
SOMETIMES WE'LL
USE DOTS.

WHAT? WHO WANTS
TO EAT DOTS?

Looking at each picture, first write the multiplication sentence it describes, and then answer the division problem. I'll do the first one for you!

1.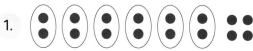

What multiplication sentence does this show?

And 16 ÷ 6 = _?_

Let's Play: Hmm, we see 6 groups, with 2 dots each. That looks like 6 × 2, right? But since we also have 4 extra dots, the full sentence would be: 6 × 2 + 4 = 16. Great! And we can figure out 16 ÷ 6, because there are 2 dots in each group and 4 left over—in other words: 16 ÷ 6 = 2 R4. Done!

Answer: 6 × 2 + 4 = 16 (or 2 × 6 + 4 = 16)
and 16 ÷ 6 = 2 R4

2.

What multiplication sentence does this show? And 9 ÷ 2 = _?_

3.

What multiplication sentence does this show? And 13 ÷ 3 = _?_

4.

What multiplication sentence does this show? And 11 ÷ 3 = _?_

5.

What multiplication sentence does this show? And 10 ÷ 4 = _?_

6.

What multiplication sentence does this show? And 13 ÷ 5 = _?_

7.

What multiplication sentence does this show? And 19 ÷ 6 = _?_

(Answers on page 222.)

Multiples—and Whipped Cream!
Bigger Division

Back in Chapter 4, we learned that for a division problem like 42 ÷ 6 = ?, all we have to do is say to ourselves, "Hmm, 6 times *what* equals 42?" Then we can remember the multiplication fact 6 × 7 = 42, and we're done, because that means 42 ÷ 6 = 7, right? But when we see a division problem that isn't part of a fact family, we have to think a little bit harder and remember our *multiples*—they're going to save the day! But first, can we talk about whipped cream for a minute?

YES.

Imagine we're making a banana split sundae, and we LOVE whipped cream. Let's use a spoon—could be big or small—and figure out the biggest number of scoops we can put on it before it all slides off. Sometimes the only way to find out how many we can *fit* is to first find out how many scoops is *too many.* So if we put 4 scoops on and then it all falls off, we know that 3 scoops is the most that can fit, right?

Let's see how many scoops will fit!

Oops! Too many scoops!

Ah . . . so 3 is the biggest number of scoops that will fit!

THAT IS SO TRUE! BUT WHAT DOES THIS HAVE TO DO WITH DIVISION?

WHO CARES?

Back on p. 34, we saw how to find multiples of a number on the multiplication chart. For example, some multiples of 6 are 6, 12, 18, 24, 30, 36, 42, 48. . . . Another great way to find multiples is by thinking "6 × 1 = **6**, 6 × 2 = **12**, 6 × 3 = **18**. . . ." Now I'll show you how multiples (and whipped cream!) help us with remainder division.

×	1	2	3	4	5	6	7	8	9	10	11	12
1	**1**	2	3	4	5	6	7	8	9	10	11	12
2	2	**4**	6	8	10	12	14	16	18	20	22	24
3	3	6	**9**	12	15	18	21	24	27	30	33	36
4	4	8	12	**16**	20	24	28	32	36	40	44	48
5	5	10	15	20	**25**	30	35	40	45	50	55	60
6	6	12	18	24	30	**36**	42	48	54	60	66	72
7	7	14	21	28	35	42	**49**	56	63	70	77	84
8	8	16	24	32	40	48	56	**64**	72	80	88	96
9	9	18	27	36	45	54	63	72	**81**	90	99	108
10	10	20	30	40	50	60	70	80	90	**100**	110	120
11	11	22	33	44	55	66	77	88	99	110	**121**	132
12	12	24	36	48	60	72	84	96	108	120	132	**144**

Let's do 29 ÷ 6 = ? We could say to ourselves, "How many times does 6 fit inside 29?" In other words, "How many times does 6 *go into* 29?" We need to check multiples of 6 near 29 to <u>find the biggest multiple of 6 *smaller* than 29</u>. Let's think of a multiple of 6 that is a little bit smaller than 29. Hmm, 6 × 3 = 18, and 6 × 4 = 24, which is pretty close to 29! But is 24 the biggest multiple of 6 smaller than 29? Hmm . . . well, let's try 6 × 5 = 30. Oops! Since 30 is bigger than 29, that was too many 6's! So we should stick to the 4 scoops: <u>6 × **4** = 24</u>.

Let's see how many 6's will fit!　　Oops! Too many 6's!　　Ah . . . so 4 is the biggest number of 6's that will fit!

Notice that because 4 × 6 only equals 24, there's still a little room on the "29" sundae. (Just not enough for an entire "6 scoop"!) Since 29 − 24 = 5, we get a remainder of **5**. Yep, 29 ÷ 6 = **4 R5**.

The remainder will always be smaller than the divisor—otherwise, we picked the wrong multiple! For example, for $65 \div 7 = ?$, here are two ways it could go:

WRONG	RIGHT
Hmm, $65 \div 7 = ?$	Hmm, $65 \div 7 = ?$
Let's check some multiples of 7: $7 \times 7 = 49$, $7 \times 8 = 56$. Oh, that's pretty close to 65 without being too big. I don't feel like checking anymore, and I'm sure this 56 is fine. So since $7 \times 8 = 56$, that means 7 goes into 65, 8 times, and what's the remainder? $65 - 56 = 9$.	Let's check some multiples of 7 that are near 65: $7 \times 7 = 49$, $7 \times 8 = 56$, $7 \times 9 = 63$, $7 \times 10 = 70$. Okay, so 70 is too big (whipped cream fell off!), which means 63 is the right multiple. So since $7 \times 9 = 63$, that means 7 goes into 65, yep, 9 times, and what's the remainder? $65 - 63 = 2$.
$65 \div 7 = $ **8 R9**	
NOPE!	CORRECT ANSWER:
The remainder cannot be bigger than the divisor—which here is 7!	$65 \div 7 = $ **9 R2**

Be sure to check all the multiples near the total until you find the one that's too big, and then you'll know you didn't miss anything!

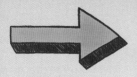

GAME TIME!

Do these division problems with remainders. I'll do the first one for you!

1. 77 ÷ 12 = __?__

Whipped Cream

Let's Play: So, "How many times does 12 go into 77?" In other words, "How many scoops of 12 fit inside 77?" To get that answer, we'll start listing multiples of 12 that are less than 77. Hmm, 12 × 5 = 60, 12 × 6 = 72, 12 × 7 = 84. Oops! 84 is too big (7 is too many scoops of 12!), which means we want to use 12 × 6 = __72__. Great! So 12 fits inside 77 a total of 6 times, and how much is left over? Well, 77 − 72 = 5, so we have **5** left over, which will be our remainder. Done!

Answer: 77 ÷ 12 = 6 R5

2. 24 ÷ 5 = __?__

3. 19 ÷ 3 = __?__

4. 27 ÷ 4 = __?__

5. 37 ÷ 9 = __?__

6. 75 ÷ 8 = __?__

7. 35 ÷ 3 = __?__

8. 44 ÷ 6 = __?__

9. 53 ÷ 7 = __?__

10. 32 ÷ 5 = __?__

11. 41 ÷ 10 = __?__

12. 87 ÷ 8 = __?__

13. 62 ÷ 5 = __?__

14. 89 ÷ 11 = __?__

15. 26 ÷ 9 = __?__

16. 52 ÷ 6 = __?__

17. 145 ÷ 12 = __?__

(Answers on page 222.)

Chapter 8

More Bagels and Royal Kitties: Multi-Digit Multiplication

By now, we've had lots of practice with our multiplication facts 0 to 12, so let's talk about some ways to multiply bigger numbers.

WHOA, WHOA--I CAME HERE FOR THE BAGELS. I'M NOT READY TO LEARN SOMETHING NEW.

ACTUALLY, IT'S REALLY NOT NEW-- YOU'VE BEEN DOING IT WITH THE 12'S THIS WHOLE TIME!

Return of the Bagels . . . and the Distributive Property

We first learned the Distributive Property on p. 41, which can turn a big multiplication problem into two smaller, easier ones, and we used it for our 12's multiplication facts. We can think about bagels or hammers when we split numbers like we did with 12×7 on p. 129. But this time, let's switch the order to 7×12 and I'll show you how to use parentheses.

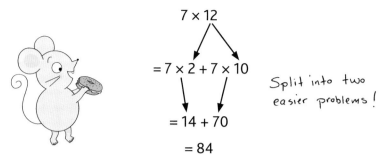

$$7 \times 12$$
$$= 7 \times 2 + 7 \times 10$$
Split into two easier problems!
$$= 14 + 70$$
$$= 84$$

There's another step we can add when we use the Distributive Property, and it will be helpful when the numbers get bigger. So let's do that same problem again, 7×12, with the added step. This time, when we split the bagel (the 12), we'll leave it where it was and put it in parentheses:

$$7 \times 12 = 7 \times (10 + 2)$$

All we did is rewrite the 12 as 10 + 2, see? And the parentheses are important—because otherwise it might look like the 2 wasn't part of the multiplication! With me so far? Great! Now that it's written like this, we can imagine the 7 is going to put honey on *both parts* of the 12:

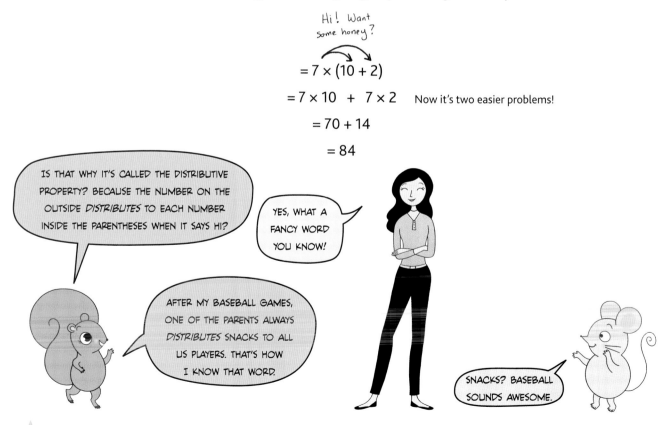

Hi! Want some honey?

$$= 7 \times (10 + 2)$$
$$= 7 \times 10 \; + \; 7 \times 2 \quad \text{Now it's two easier problems!}$$
$$= 70 + 14$$
$$= 84$$

IS THAT WHY IT'S CALLED THE DISTRIBUTIVE PROPERTY? BECAUSE THE NUMBER ON THE OUTSIDE *DISTRIBUTES* TO EACH NUMBER INSIDE THE PARENTHESES WHEN IT SAYS HI?

YES, WHAT A FANCY WORD YOU KNOW!

AFTER MY BASEBALL GAMES, ONE OF THE PARENTS ALWAYS *DISTRIBUTES* SNACKS TO ALL US PLAYERS. THAT'S HOW I KNOW THAT WORD.

SNACKS? BASEBALL SOUNDS AWESOME.

When using the Distributive Property, make sure there's *addition* in the parentheses, not multiplication. For example, if we have 5 × (10 × 3), we just multiply everything together—there's <u>nothing to distribute</u>.

OK	NO	YES
(We don't need the Distributive Property! It's all just multiplication.)	(Seriously, we cannot use the Distributive Property!)	(This is a <u>different problem</u>, and a correct use of the Distributive Property.)

OK

5 × (10 × 3) = ?

= 5 × (30)

= 5 × 30

= **150**

NO

5 × (10 × 3) = ?

This is wrong!

= 5 × (10 × 3)=

5 × 10 + 5 × 3

= 50 + 15

= 65   *Huh? Nope!*

YES

5 × 13 = ?

= 5 × (10 + 3)

= 5 × 10 + 5 × 3

= 50 + 15

= **65**

OH, THAT MAKES SENSE! IF IT'S ONLY MULTIPLICATION AND WE DON'T SPLIT ANYTHING WITH ADDITION, THEN IT WOULDN'T MAKE SENSE TO *ADD* ANYTHING BACK TOGETHER.

THAT'S A GREAT WAY TO THINK ABOUT IT, MS. SQUIRREL!

QUICK NOTE

The full, fancy name is the *Distributive Property of Multiplication over Addition*, and now you understand why: If there's no addition in the parentheses, we can't distribute!

Stretching Kitties . . . and Expanded Form

Cats are great, aren't they? And they're so cute when they stretch. I mean, check it out: One minute, she's this cute little kitty, and the next minute she's looooong. It's the same cat, stretched out! We can do the same thing with numbers. It's called *expanded form.* You've probably seen it before, but let's review.

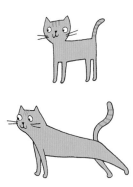

??? WHAT'S IT CALLED? ???

Writing a number in **expanded form** means splitting it up to show how much *each digit stands for.* For example:

28 in *expanded form* is 20 + 8.

63 in *expanded form* is 60 + 3.

425 in *expanded form* is 400 + 20 + 5.

It's the *same number,* stretched out!

Standard Form: She's just *standing*, after all!	Expanded Form: Same cat, stretched and *expanded*!
425	400 + 20 + 5

See p. 104 in *Do Not Open This Math Book* for more on this!

And what does this have to do with multiplication? Well, when we split up numbers to use the Distributive Property, it's really useful to split them up into their *expanded forms.* Let's see how this works with 3 × 24 = ? It might seem a little scary at first, but it's really not so bad—we'll use expanded form, and then use the Distributive Property like we did with 7 × 12 on p. 164. Let's do it!

$3 \times 24 = ?$

$= 3 \times (20 + 4)$ We wrote 24 in expanded form!

$= 3 \times (20 + 4)$ The Distributive Property!

$= 3 \times 20 \ + \ 3 \times 4$

$= 60 + 12$

$= \mathbf{72}$

> **QUICK NOTE**
>
> Many of these problems (like #1 in the Game Time on the next page) will have bigger numbers in them that will get added together, so if you'd like to review *addition with regrouping,* check out Chapter 9 of my addition & subtraction book *Do Not Open This Math Book.* (Mr. Mouse is happy to meet you there. Really.)

GAME TIME!

Do these multiplication problems by first writing numbers in expanded form, distributing, solving the easier multiplication problems, and then adding them together for the answer. I'll do the first one for you!

1. $8 \times 239 =$ _?_

Let's Play: Yikes, this looks hard, but we won't be scared off! First, we'll rewrite the problem with 239 stretched into expanded form: $8 \times (200 + 30 + 9) = ?$ Then we'll distribute the 8 to the 200, 30, and 9:

$= 8 \times (200 + 30 + 9)$

The Distributive Property!

$= 8 \times 200 + 8 \times 30 + 8 \times 9$

Now we'll solve the "easier" multiplication problems: First let's do 8×200. Since $8 \times 2 = 16$, we know $8 \times 200 = \underline{1,600}$. (See p. 151!) Next, $8 \times 30 = \underline{240}$, and $8 \times 9 = \underline{72}$, so we just add:

$1,600 + 240 + 72 = 1,912.$

$$\begin{array}{r} \overset{1}{1600} \\ 240 \\ +\ \ \ 72 \\ \hline 1912 \end{array}$$

And that's our answer: 1,912. Done!

Answer: $8 \times 239 = 1,912$

2. $3 \times 46 =$ _?_

3. $5 \times 19 =$ _?_

4. $4 \times 63 =$ _?_

5. $2 \times 89 =$ _?_

6. $7 \times 26 =$ _?_

7. $6 \times 82 =$ _?_

8. $8 \times 88 =$ _?_

9. $4 \times 235 =$ _?_

10. $3 \times 412 =$ _?_

(Answers on page 222.)

Pretty Pictures: Area and Box/Grid Methods

Let's see some of the ways the (kitty stretch!) expanded form of multiplication can look—you might see some of these at your school! We'll do the example 5 × 26 in three different ways. For all of them, we'll start by writing 26 in expanded form, 20 + 6, and then use the Distributive Property to do the multiplication.

Arrays of Dots

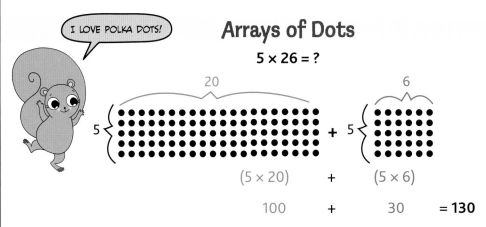

I LOVE POLKA DOTS!

5 × 26 = ?

20   6

5   +   5

(5 × 20)   +   (5 × 6)

100   +   30   = **130**

This might remind you of when we split up the bagels on p. 41!

Area Models

5 × 26 = ?

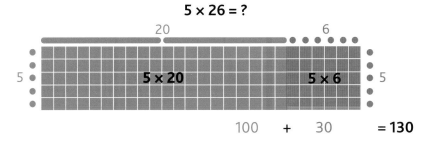

20   6

5   **5 × 20**   **5 × 6**   5

100   +   30   = **130**

These might also remind you of the hammer from p. 43! But notice that this time, there's a grid inside the rectangles. That shows *area*. We'll do more with *area models* in Chapter 9.

Box/Grid—Rough and *Not* So Pretty!

$$5 \times 26 = ?$$

        20   6
    ┌────────┬──────┐
  5 │ 100   │ 30 │
    └────────┴──────┘
      100 + 30 = 130

That's a rough sketch of how we can break up the multiplication into two parts.
You might be asked to do this on your homework sometimes!

Itty-Bitty Kitty Stretch: Partial Products

Sometimes a kitty might do a mini-stretch, like this, where her back arches up just a little bit.
You can hardly tell she's stretching at all!

SHE'S SO CUTE, DOING HER LITTLE SECRET STRETCH. . . .

LIKE A HALLOWEEN HORROR MOVIE POSTER!

Just like how kitties don't always stretch out, we won't always want to write the numbers in expanded form for multiplication problems—especially as the numbers get bigger. (It'll also be much easier to leave numbers in standard form when we stack the factors on top of each other, which we'll see in a moment.)

But even without *writing* the numbers all the way out in expanded form, we can still multiply the expanded parts separately. It's a lot like the kitty who still looks the same size . . . but we know she's really stretching. Many call this the *partial products* method, because we'll multiply the numbers in <u>parts</u>. Let's do one:

$$\begin{array}{r} 79 \\ \times\ 6 \\ \hline \end{array}$$

Ooh, this looks challenging—but we can do it! Notice that the ones digit of 79—the 9—is directly over the 6. It's always important to keep our numbers lined up correctly!

We'll multiply 6 times the 79 in two <u>parts</u> below: First, we'll multiply the 6 times the 9, write that answer (54), then multiply the 6 times the 70 (remember it's a 70, not just a "7"). We get 6 × 70 = 420, then write that below, and then add the two "partial products" together for the total answer: 54 + 420 = **474.** Ta-da!

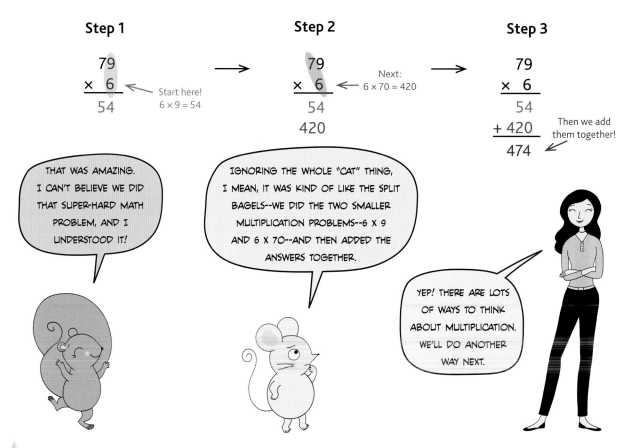

Step 1

$$\begin{array}{r} 79 \\ \times\ 6 \\ \hline 54 \end{array}$$

Start here!
6 × 9 = 54

Step 2

$$\begin{array}{r} 79 \\ \times\ 6 \\ \hline 54 \\ 420 \end{array}$$

Next:
6 × 70 = 420

Step 3

$$\begin{array}{r} 79 \\ \times\ 6 \\ \hline 54 \\ +\ 420 \\ \hline 474 \end{array}$$

Then we add them together!

THAT WAS AMAZING. I CAN'T BELIEVE WE DID THAT SUPER-HARD MATH PROBLEM, AND I UNDERSTOOD IT!

IGNORING THE WHOLE "CAT" THING, I MEAN, IT WAS KIND OF LIKE THE SPLIT BAGELS--WE DID THE TWO SMALLER MULTIPLICATION PROBLEMS--6 X 9 AND 6 X 70--AND THEN ADDED THE ANSWERS TOGETHER.

YEP! THERE ARE LOTS OF WAYS TO THINK ABOUT MULTIPLICATION. WE'LL DO ANOTHER WAY NEXT.

In the (itty-bitty kitty) partial products method, <u>we have to think about place value</u>. For example, in the problem we just did, when we multiplied the 6 times the 7, we were really multiplying the 6 times 70, <u>and we had to remember that</u>, so we could correctly write 420.

The itty-bitty kitty partial products method is nice because we can really see what's going on. But the place value stuff can be tricky to remember, so here's another way!

Cowboy Kitty Roping:
The Traditional Method

In the itty-bitty kitty stretch (partial products method) we saw on p. 172, even though we didn't write the factors in expanded form, the answer was still "expanded out" and we had to add it back together to get our answer: 54 + 420 = 474. In this next method, we won't have to add anything at the end. Yep, the kitty stretch is totally invisible. This time, she's only *thinking* about stretching. Oh, and she's a cowboy.

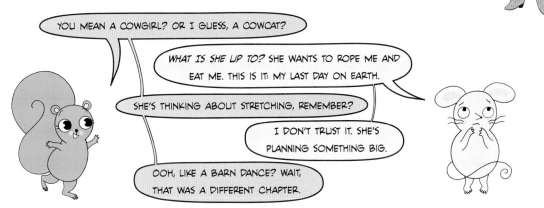

YOU MEAN A COWGIRL? OR I GUESS, A COWCAT?

WHAT IS SHE UP TO? SHE WANTS TO ROPE ME AND EAT ME. THIS IS IT: MY LAST DAY ON EARTH.

SHE'S THINKING ABOUT STRETCHING, REMEMBER?

I DON'T TRUST IT. SHE'S PLANNING SOMETHING BIG.

OOH, LIKE A BARN DANCE? WAIT, THAT WAS A DIFFERENT CHAPTER.

Cowboy Kitty doesn't mess around—she ropes her numbers and gets the answer fast! For 26 × 3, let's imagine that the 3 is Cowboy Kitty standing on the ground, and she's going to toss her lasso high into the air and rope the digits of 26, one at a time. She always starts with the ones digits, so first, she ropes the 6, and we get 3 × 6 = 18.

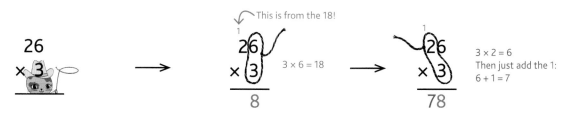

This is from the 18!

$3 \times 6 = 18$

$3 \times 2 = 6$
Then just add the 1:
$6 + 1 = 7$

If we were doing the partial products method, we'd write the entire 18 under the line, but Cowboy Kitty doesn't have time for that, so we just write the 8 below the line, and we "regroup" the 1 to the top of the *tens* column, kind of like we learned with addition (see p. 119 in *Do Not Open This Math Book*). Because, after all, it's really a 10, so the tens column is where it belongs! Next, look at the problem on the right above to see how our "3 kitty" ropes the 2, so 3 × 2 = 6. Now we just add the 1 we regrouped to get 6 + 1 = 7. And there's our answer! 26 × 3 = 78. Ta-da!

Next, let's see how the traditional "Cowboy Kitty" method works on the 79 × 6 problem from p. 172. She keeps it fast and regroups the tiny numbers on top to get the job done! Let's imagine the 6 is Cowboy Kitty. First we rope the 9 and get 6 × 9 = 54, then we write the 4 below the line and "regroup" the 5 to the top of the tens column.

This is the 5 from the 54!

$6 \times 9 = 54$

$6 \times 7 = 42$
Then just add the 5:
$42 + 5 = 47$

Write this here!

Then the kitty ropes the 7, we get 6 × 7 = 42, and finally we __add__ the 5 we regrouped, to get 42 + 5 = 47. We write the 47 below to get **474**, and we're done: 79 × 6 = 474!

Once you get good at the traditional "Cowboy Kitty" method, there won't be much to write down. You'll do most of it in your head, *imagining* the ropes, and it'll look something like this:

Cowboy Kitty would be so proud.

$$\begin{array}{r} ^{5} \\ 79 \\ \times\ 6 \\ \hline 474 \end{array}$$

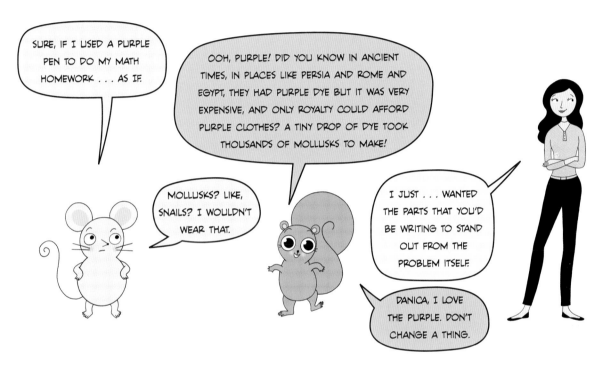

SURE, IF I USED A PURPLE PEN TO DO MY MATH HOMEWORK . . . AS IF.

OOH, PURPLE! DID YOU KNOW IN ANCIENT TIMES, IN PLACES LIKE PERSIA AND ROME AND EGYPT, THEY HAD PURPLE DYE BUT IT WAS VERY EXPENSIVE, AND ONLY ROYALTY COULD AFFORD PURPLE CLOTHES? A TINY DROP OF DYE TOOK THOUSANDS OF MOLLUSKS TO MAKE!

MOLLUSKS? LIKE, SNAILS? I WOULDN'T WEAR THAT.

I JUST . . . WANTED THE PARTS THAT YOU'D BE WRITING TO STAND OUT FROM THE PROBLEM ITSELF.

DANICA, I LOVE THE PURPLE. DON'T CHANGE A THING.

Just for fun, let's do that same problem using a box picture: 6 × 79 = ? We'll write 79 in expanded kitty stretch form: 70 + 9, draw our boxes, multiply each rectangle (6 × 70 = 420 and 6 × 9 = 54), and then add 'em up!

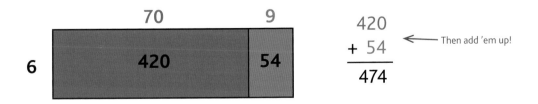

Then add 'em up!

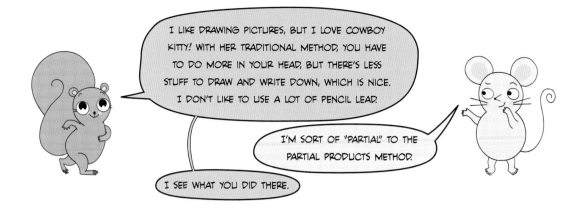

I LIKE DRAWING PICTURES, BUT I LOVE COWBOY KITTY! WITH HER TRADITIONAL METHOD, YOU HAVE TO DO MORE IN YOUR HEAD, BUT THERE'S LESS STUFF TO DRAW AND WRITE DOWN, WHICH IS NICE. I DON'T LIKE TO USE A LOT OF PENCIL LEAD.

I'M SORT OF "PARTIAL" TO THE PARTIAL PRODUCTS METHOD.

I SEE WHAT YOU DID THERE.

Why the "Zeros Trick" Works

To multiply any whole number times 10, we just add a zero, right? But if we didn't know the trick, we could use our Cowboy Kitty method and get the right answer, of course! Let's look at 4×10 and also the example from p. 151, 6×400:

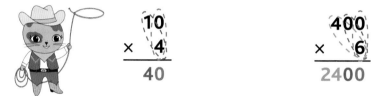

$$\begin{array}{r} 10 \\ \times\ 4 \\ \hline 40 \end{array} \qquad \begin{array}{r} 400 \\ \times\ 6 \\ \hline 2400 \end{array}$$

For the first problem, Cowboy Kitty starts by roping and multiplying $4 \times 0 = 0$, right? So we write 0 below and . . . there's nothing to regroup up on top of the 1, is there? Then we rope and multiply $4 \times 1 = 4$, which we write in the tens place, to get 40 as our answer. Same with the next problem—*we keep getting 0* when we rope. Try that one on your own!

So whether we use Cowboy Kitty or the Zeros trick from p. 151, we end up moving the *place value* of the non-zero part of the product by the *same number of places* as we have zeros. (Try reading that a few times until it makes sense to you. Nice job!)

There are lots of ways to think about multiplication—do whichever ones you like the most!

GAME TIME!

Do these multiplication problems, using the pictures from p. 170, the partial products method shown on p. 172, or the traditional (Cowboy Kitty) method from p. 174. I'll do the first one for you!

1. $584 \times 3 =$ _?_

Let's Play: Okay, deep breath. We can do this! Let's do the partial products method. We'll start by writing the numbers on top of each other. Then, remembering place value just like we did on p. 172, we'll do the partial products: $3 \times 4 = 12$, and then $3 \times 80 = 240$, and they get written below the line. Next, we have to multiply the 3 times that 5—which is really 500, since it's in the hundreds column. And $3 \times 500 = 1500$.

```
    584
  ×   3
     12
    240
   1500
```

So we put that below, and then *add up* all the partial products to get 1752. To see this same problem done with pictures and also with the traditional method, see p. 178! (For our final answer here, we'll need to add a comma. See p. 152 for more on this.)

```
     12
    240
  + 1500
  ──────
   1752
```

Answer: $584 \times 3 = 1,752$

2. $32 \times 4 =$ _?_

3. $51 \times 8 =$ _?_

4. $77 \times 7 =$ _?_

5. $68 \times 3 =$ _?_

6. $99 \times 2 =$ _?_

7. $22 \times 9 =$ _?_

8. $345 \times 2 =$ _?_

9. $411 \times 6 =$ _?_

10. $123 \times 5 =$ _?_

11. $389 \times 8 =$ _?_

12. $103 \times 4 =$ _?_

Hint: See p. 179.

13. $505 \times 5 =$ _?_

Hint: See p. 179.

How Much of a Stretch Do You Like?

It's nice to have options, right? I want you to see how #1 from the Game Time on p. 177 looks as a box picture and also in the traditional method, since it has three digits:

Box Picture Method

So for 584 × 3 = ?, we'd split up the 584 into its expanded form, 500 + 80 + 4, draw this box, multiply each of the rectangles, and then add 'em up!

	500	80	4
3	1500	240	12

$$1500$$
$$240$$
$$+ \ 12 \quad \longleftarrow \text{Then add 'em up!}$$
$$\overline{1752}$$

Cowboy Kitty "Traditional" Method

Since we have a three-digit number, I'll show you the "ropes" to more easily remember what's going on.

$$\overset{2 \ 1}{584}$$
$$\times \quad 3$$
$$\overline{1752}$$

At first, use whichever method makes the most sense to you. With a little practice, you'll be able to do them all. You got this!

Zeros in the Middle?

Sometimes we'll see a zero in the middle of a factor, if it has three or more digits. But don't worry—that actually makes the multiplication easier!

I LIKE "EASIER."

Let's do **3 × 609 = ?** First we write 609 in expanded form: 600 + 9, then draw our box, multiply each rectangle (3 × 600 = 1800 and 3 × 9 = 27) and add 'em up!

	600	9
3	1800	27

$$\begin{array}{r} 1800 \\ + \ \ 27 \\ \hline 1827 \end{array}$$

The partial products method would look like this:

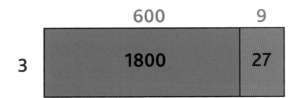

$$\begin{array}{r} 609 \\ \times \ \ \ 3 \\ \hline 27 \end{array} \rightarrow \begin{array}{r} 609 \\ \times \ \ \ 3 \\ \hline 27 \\ 0 \end{array} \rightarrow \begin{array}{r} 609 \\ \times \ \ \ 3 \\ \hline 27 \\ 0 \\ 1800 \end{array}$$

Then add 'em up!

$$\begin{array}{r} 27 \\ 0 \\ + \ 1800 \\ \hline 1827 \end{array}$$

See? There's less to add up, so it's actually easier! And here's how it looks with the traditional (Cowboy Kitty roping!) method—*without* showing the ropes this time:

$$\begin{array}{r} \overset{2}{6}09 \\ \times \ \ \ \ 3 \\ \hline 1827 \end{array}$$

Remember, if you don't like the traditional method, you can just stick to the pictures or partial products for now.

Chapter 9

Chocolate Bars and Window Breaking:
Area and Two-by-Two Multiplication

Big Bars of Chocolate—and Area!

In this chapter, we'll learn to multiply things like 19 × 24 = ? But first, we'll talk about area . . . and chocolate.

YES!

    Let's say we have a bunch of chocolate chips, and we arrange them like an array. If we know there are 4 rows and 6 columns, then we know there are 24 chips, right? And same goes if we have a chocolate bar with squares in 4 rows and 6 columns—we'd have 24 squares total.

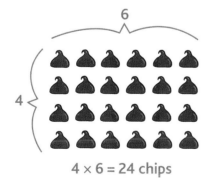

4 × 6 = 24 chips

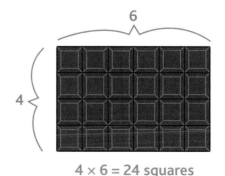

4 × 6 = 24 squares

When arrays are made of squares, they show *area*.

Area is the <u>size</u> of a two-dimensional shape, like a square or rectangle, even a circle or triangle—or something else entirely! The shapes below all have the same area of 9. They are each made from 9 squares:

If a shape is a square or rectangle, we can find the *area* by multiplying the length times the width (the rows times the columns).

Area = length × width

In the chocolate bar example on p. 180, the area is 24, because 4 × 6 = 24. Yep, when arrays are made of squares, they show area.

I LIKE THINKING ABOUT MULTIPLICATION WITH THE CHOCOLATE CHIPS--THEY ARE SO CUTE!

I MELT MY CHOCOLATE AND SPREAD IT ON GRAHAM CRACKERS, SO I LIKE *AREA* S'MORE . . . GET IT?

YUMMY! BUT HOW WOULD I KEEP TRACK OF *HOW MUCH* CHOCOLATE I'M SPREADING, YOU KNOW, TO STAY HEALTHY AND NOT OVERDO IT?

"OVERDO"? I DON'T UNDERSTAND.

WE'LL TALK ABOUT UNITS OF AREA TO MEASURE THAT CHOCOLATE, BUT FIRST, LET'S MAKE SURE WE KNOW WHAT *UNITS* ARE!

??? WHAT'S IT CALLED? ???

Units are types of measurements. For example, some units of time are seconds, hours, and years. We can measure weight with units like pounds and kilograms, and we can measure volume with units like gallons and liters. Some units of length are inches, centimeters, and miles. Units can measure area, too, like we'll see below. And there are many more types of units!

Units of Area

If the <u>length</u> of an object is measured in **inches,** then its <u>area</u> can be measured in **square inches,** which can look like this: **in².** Same goes for centimeters (cm), feet (ft), kilometers (km), miles (mi), and more—when it's area, we add the word "square," or add a little 2 up above, like **square kilometers,** or **km².** See what I mean?

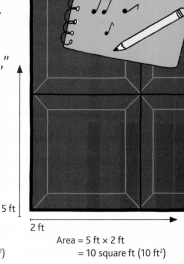

QUICK NOTE

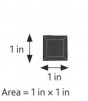

1 in

1 in

Area = 1 in × 1 in
= 1 square inch (1 in²)

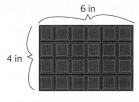

6 in

4 in

Area = 4 in × 6 in
= 24 square inches (24 in²)

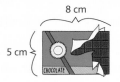

8 cm

5 cm

Area = 5 cm × 8 cm
= 40 square cm (40 cm²)

5 ft

2 ft

Area = 5 ft × 2 ft
= 10 square ft (10 ft²)

OH, THAT LITTLE "CENTIMETER" CHOCOLATE BAR IS SO TINY AND CUTE!

GIMME.

Multiply the two sides to find the area of these rectangles. Pretend these numbers mean chocolate chips or little squares, if that helps! Be sure to write the square unit in your answer: square feet, square inches, etc. Or use abbreviations like ft² and in². I'll do the first one for you!

1.

50 km

20 km

Let's Play: Okay, we're being asked to find the area of this blue-green rectangle. If we imagine we have an array with 20 rows and 50 columns of chocolate chips, then we know we could just multiply those two numbers together to get the total number of chocolate chips, right? This works for area too:

Area = 20 × 50

We know from the zeros trick on p. 151 how to do 20 × 50 = 1,000. But we're not done! Since the <u>lengths</u> of the sides were labeled in kilometers, that means the shape's <u>area</u> is measured in *square* kilometers, or km² for short. So our answer is 1,000 square kilometers (or 1,000 km²). Done!

Answer: 1,000 square kilometers, or 1,000 km²

2.

9 cm

6 cm

3.

12 m

9 m

4.

20 mi

10 mi

5.

7 ft

7 ft

6.

3 in

7 in

7.

12 mm

30 mm

(Answers on page 223.)

Grow Your Own Food: Area Models for Two-by-Two Multiplication

Yep, the Aztecs were great at farming, and farming is going to help us to multiply two-digit numbers. This will be like the area models we saw on p. 183. Let's do it!

To multiply **13 × 15,** imagine we have a rectangular area of farmland 13 km long and 15 km wide. If we could multiply 13 × 15, we'd get the whole area, right? But what if we don't know *how* to multiply 13 × 15?

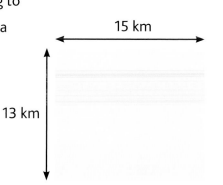

Let's pretend we're Aztecs. We'll split up the land into four sections for growing corn, tomatoes, avocados . . . and just so Mr. Mouse will be happy, cacao trees (you know, for making chocolate). If we split up 13 and 15 in (stretched-out kitty) *expanded form* and break up the big rectangle like that, then our farm might look something like this:

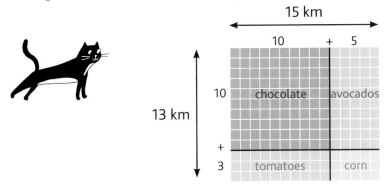

Give this a careful look: It's the same area, but broken up into four parts. Notice that the way the lines break this up, the tomato area is 3 km high and 10 km wide—in other words, 3 × 10. And do you see how the avocado area is 10 × 5? Can you figure out the area for the corn and chocolate, too?

By splitting up the farm, we break up 13 × 15 into four easier multiplication problems!

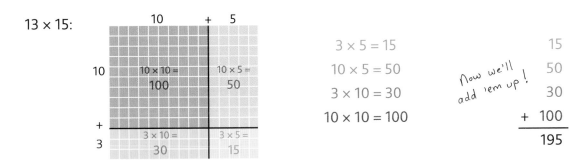

13 × 15:

$3 \times 5 = 15$

$10 \times 5 = 50$

$3 \times 10 = 30$

$10 \times 10 = 100$

Now we'll add 'em up!

15
50
30
+ 100
―――
195

By adding up all four smaller areas, we found the total area: 13 × 15 = **195**. And if you counted all the little squares in the farm diagram on the previous page, yep, you'd get 195. Our answer should use the correct units, square kilometers: 13 km × 15 km = **195 km²**. Done!

Here's how it might look on homework:

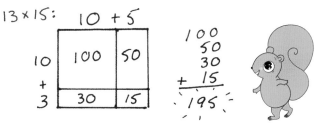

I think Ms. Squirrel added those sparkles—your teacher might prefer a *box* around your answer. Speaking of boxes, let's learn the window (or "box" or "grid") method!

Don't Break the Glass Window!
Multiplication with the Box (or Grid) Method

In addition to area models, another great way to do multiplication of two-digit numbers is the box, or grid, method. But I like to call it the *window* method, and I think it's "clear" why. . . .

Let's do 79 × 28 = ? Even though 79 is much bigger than 28, in the window method, it doesn't matter! We can just draw a regular box, split it into four squares like a window, and write the kitty-stretched *expanded form* of the factors along the sides.

79 × 28:

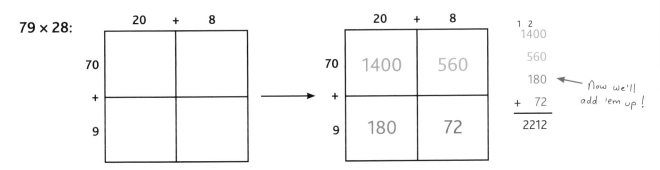

Then we multiply to get all four products, write them in the boxes, and add up the products to get our answer: 1400 + 560 + 180 + 72 = **2,212.** Notice the little 1 and 2 that get regrouped up top. To review *addition with regrouping,* check out Chapter 9 in *Do Not Open This Math Book.*

> **QUICK NOTE**
>
> Many teachers don't include the plus signs outside the window (or box), and some even put *extra* boxes around it and include a × symbol. So you could see windows that look like this, but don't worry—they work exactly the same way.
>
> | | 20 | 8 |
> |----|------|-----|
> | 70 | 1400 | 560 |
> | 9 | 180 | 72 |
>
> Two other ways it could look!
>
> | × | 20 | 8 |
> |----|------|-----|
> | 70 | 1400 | 560 |
> | 9 | 180 | 72 |

STEP BY STEP:

Multiplying Two-Digit Numbers with Window (Box/Grid) Models

Step 1. Draw a "window" with four squares.

Step 2. Write both factors in their expanded form on the top and left side.

Step 3. Do all four multiplication problems and write those answers inside the little boxes.

Step 4. Add up the four products for the total answer!

By the way, the only difference between the window (box/grid) model and the area model is the size of the boxes—you can actually use the above Step by Step for both!

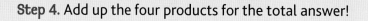

Area Model for 86 × 34	Window (box/grid) model for 86 × 34
80 + 6 30 2400 180 + 4 320 24	80 + 6 30 2400 180 + 4 320 24

In both cases, we do the four multiplication problems, and then add up the products for our answer: in this case, 2400 + 180 + 320 + 24 = **2,924**. So 86 × 34 = 2,924. Ta-da!

 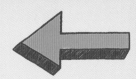

GAME TIME!

Multiply using the area or box (window!) method, whichever you prefer.
For the first few, you'll fill in blanks in the windows, and then solve it.
After that, you'll do the entire thing on your own. I'll do the first one for you!

1. 13 × 15 = _?_

Let's Play: This is the farm problem from p. 184. This time, we'll do the window (box/grid) method! We'll draw a box and write the 13 and 15 in expanded form on the sides: 10 + 3 and 10 + 5. Next, we'll do the four multiplication problems, put them in the boxes, and then add 'em up: 100 + 50 + 30 + 15 = 195.

I've shown you in my writing so you can see how it might look on your homework. To see this problem using the area model—also in my writing—check out p. 186!

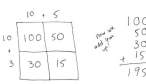

Answer: 13 × 15 = 195

2. 45 × 16
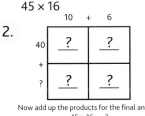
Now add up the products for the final answer.
45 × 16 = _?_

3. 23 × 52
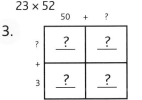
Now add up the products for the final answer.
23 × 52 = _?_

4. 31 × 86
Now add up the products for the final answer.
31 × 86 = _?_

5. 67 × 22 = _?_

6. 78 × 12 = _?_

7. 43 × 64 = _?_

8. 56 × 39 = _?_

9. 48 × 92 = _?_

10. 99 × 99 = _?_

SOMEONE ONCE SAID THAT "THE EYES ARE THE <u>WINDOWS</u> TO THE SOUL" . . . I LOVE THIS CHAPTER.

(Answers on page 223.)

No More Drawing: Partial Products

Once we're really comfortable with window/box multiplication, we can do it *without* the window. We'll still end up with four easier multiplication problems, and we'll still add them up to get our answer—there just won't be any windows to break . . . so you can relax, Mr. Mouse.

Let's do the problem from p. 189: 13 × 15 = ? Instead of drawing a box, we can just write the numbers on top of each other, and we could even "kitty stretch" them into expanded form. (See p. 167 for more on expanded form . . . and stretching kitties.)

$$
\begin{array}{r} 13 \\ \times\ 15 \\ \hline \end{array}
\longrightarrow
\begin{array}{r} 10 + 3 \\ \times\ 10 + 5 \\ \hline \end{array}
\longrightarrow
\begin{array}{r} 10 + 3 \\ \times\ 10 + 5 \\ \hline 15 \\ 50 \end{array}
\longrightarrow
\begin{array}{r} 10 + 3 \\ \times\ 10 + 5 \\ \hline 15 \\ 50 \\ 30 \\ 100 \end{array}
$$

Then, add 'em up!

$$
\begin{array}{r} 15 \\ 50 \\ 30 \\ +\ 100 \\ \hline 195 \end{array}
$$

We start with the ones digits: 5 × 3 = 15. Then, staying with the 5, we do 5 × 10 = 50. Next, we move to the 10 on the bottom and do 10 × 3 = 30, and finally we multiply: 10 × 10 = 100. By breaking up the numbers, we've turned this into four easier multiplication problems—just like the four boxes we've been drawing! And just like before, we add up the products to get the final answer: 15 + 50 + 30 + 100 = **195**. Not so bad, right?

We don't have to write the numbers in "stretching kitty" expanded form—the problem on the previous page also could have looked like this (an itty-bitty kitty stretch—yep, like the partial products method from p. 172!). We just have to remember that, in this case, the 1's are really 10's! Use whichever method you like most.

Remember to start with the ones digits:

$5 \times 3 = 15$

$5 \times 10 = 50$

$10 \times 3 = 30$

$10 \times 10 = 100$

$$\begin{array}{r} 13 \\ \times\ 15 \\ \hline 15 \\ 50 \end{array}$$

→

$$\begin{array}{r} 13 \\ \times\ 15 \\ \hline 15 \\ 50 \\ 30 \\ 100 \end{array}$$

→

Then add 'em up!

$$\begin{array}{r} 15 \\ 50 \\ 30 \\ +\ 100 \\ \hline 195 \end{array}$$

IS THERE SOME WAY WE CAN DO THESE WITHOUT ANY STRETCHING CATS?!

NO MORE STRETCHING CATS.

THANK GOODNESS.

JUST A COUPLE OF COWBOY KITTIES.

GRRRRRR . . .

YEE-HAW!

Cowboy Kitty, Take Two!
Multiplying *Two* Two-Digit Numbers

Doing multiplication with pictures can really help us understand what's happening when two-digit numbers multiply together. But just like on p. 173, there is also a "traditional" way of doing this—using TWO cowboy kitties this time!

Let's try **79 × 36 = ?** Cowboy kitties stand on the ground, and they throw their ropes up into the air. And we always start with the *ones digits.* So, we'll start with the 6 kitty, and she'll throw her rope up to get the 9—we'll multiply both ones digits together: 6 × 9 = 54. By the way, this first kitty doesn't touch the 3—we could almost pretend it's not there! And that means the beginning of this problem is *exactly* the same as 79 × 6, which we did on p. 174. (Take a look if you need a refresher!)

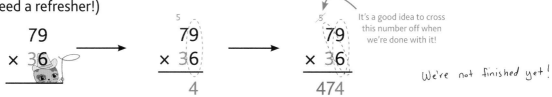

It's a good idea to cross this number off when we're done with it!

We're not finished yet!

Okay, now the 6 kitty is done multiplying, and it's time for the next kitty—the 3. Of course, it's really *30,* right? So everything we multiply will be *10 times* bigger than if it were just 3. Because of that, <u>we'll write 0 in the ones place</u> below. Next, the 3 kitty ropes the 9—starting with the ones place, as always—and we get 3 × 9 = 27. We write the 7 below, and just like before, we have to regroup a digit (this time, a 2).

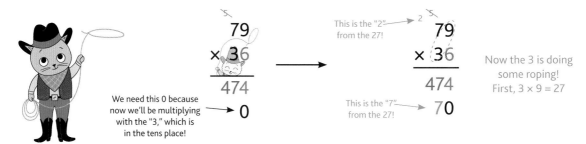

We need this 0 because now we'll be multiplying with the "3," which is in the tens place!

This is the "2" from the 27!

This is the "7" from the 27!

Now the 3 is doing some roping! First, 3 × 9 = 27

Only one roping (multiplication) left: 3 × 7 = 21, and let's not forget to add the 2 (that we regrouped) to the 21, and we get 23, which gets written below. Once we've done all four "mini" multiplication problems, we add up the products to get our answer!

The final roping!

3 × 7 = 21
21 + 2 = 23

$$
\begin{array}{r}
\overset{2}{7}9 \\
\times\ 36 \\
\hline
474 \\
2370 \\
\end{array}
$$

$$
\begin{array}{r}
\overset{2}{7}9 \\
\times\ 36 \\
\hline
\overset{1}{4}74 \\
+\ 2370 \\
\hline
2844 \\
\end{array}
$$

YEE-HAW!

Now we'll add 'em up, right here!

Ta-da! We get 79 × 36 = **2,844**. By the way, once you get good at this "traditional" method and are doing the roping in your head, it will look more like this:

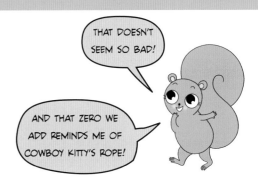

$$\begin{array}{r} 79 \\ \times\ 36 \\ \hline {}^{1}474 \\ +\ 2370 \\ \hline 2844 \end{array}$$

THAT DOESN'T SEEM SO BAD!

AND THAT ZERO WE ADD REMINDS ME OF COWBOY KITTY'S ROPE!

And here's how this same problem looks if we stretch out the number and use a window box:

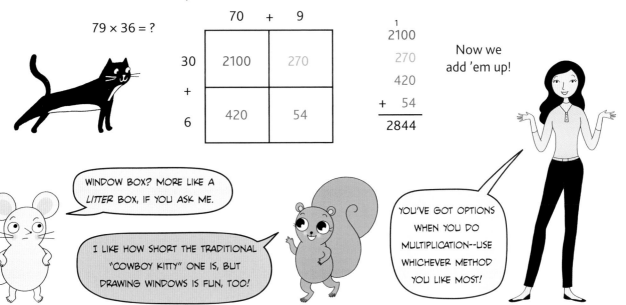

79 × 36 = ?

	70 +	9
30	2100	270
+		
6	420	54

$$\begin{array}{r} {}^{1}2100 \\ 270 \\ 420 \\ +\ 54 \\ \hline 2844 \end{array}$$

Now we add 'em up!

WINDOW BOX? MORE LIKE A *LITTER* BOX, IF YOU ASK ME.

I LIKE HOW SHORT THE TRADITIONAL "COWBOY KITTY" ONE IS, BUT DRAWING WINDOWS IS FUN, TOO!

YOU'VE GOT OPTIONS WHEN YOU DO MULTIPLICATION--USE WHICHEVER METHOD YOU LIKE MOST!

The Lattice Method

There's another wacky but cool method for multiplying called the *lattice method*, which you might see at school. If so, check it out at **TheTimesMachine.com**!

QUICK NOTE

As you get better at the traditional (Cowboy Kitty) method, you'll see it's faster than the window and partial products methods. And remember, if your teacher wants you to do the traditional method, you can always do the window method on the side of your paper to check your work!

GAME TIME!

Do these problems, using either the traditional (Cowboy Kitty), partial products, or window method. I'll do the first one for you!

1. 89 × 67 = __?__

Let's Play: We'll do partial products first. Remember, with this method, we have to keep track of what each digit really stands for, like how the 8 is really 80. Let's do it!

$$7 \times 9 = 63$$
$$7 \times 80 = 560$$

```
  89
× 67
─────
  63   ← Start here!
 560
```

Next we go here!

$$60 \times 9 = 540$$
$$60 \times 80 = 4800$$

```
  89
× 67
─────
  63
 560
 540
4800
```

Now we add 'em up!

```
    63
   560
   540
+ 4800
──────
  5963
```

Let's also do the traditional (Cowboy Kitty) method. Starting with the ones digit, first the 7 ropes both top numbers, then the 6 does. We always start with the ones digits:

YEE-HAW!

$$9 \times 7 = 63$$
$$7 \times 8 + 6 = 62$$

```
   6
  89
× 67
────
 623   ← Start here with the 7!
```

Next, we do the 6!

$$6 \times 9 = 54$$
$$6 \times 8 + 5 = 53$$

```
   5
  89
× 67
─────
  623
+5340   ← Don't forget to add the 0!
─────
 5963
```

And yep, both methods got us 5,963. Feel free to check this with the window method, too. Done!

Answer: 89 × 67 = 5,963

2. 23 × 27 = __?__

3. 51 × 31 = __?__

4. 62 × 43 = __?__

5. 18 × 19 = __?__

6. 34 × 52 = __?__

7. 72 × 12 = __?__

8. 52 × 55 = __?__

9. 67 × 22 = __?__

10. 92 × 11 = __?__

(Answers on page 223.)

Chapter 10

Fashion Models and Halloween Candy: Methods for Long Division

In this chapter, we'll learn how to divide things like 852 ÷ 6 = ? There are many ways long division is taught nowadays, but they all boil down to the same thing: splitting up a big number into even groups, chunk by chunk . . . <u>starting with the biggest chunks</u>.

But first, we're going to talk about Halloween candy. Yep, a huge pile of Halloween candy. Let's say you and five friends went trick-or-treating and made a deal that at the end of the night, you'd pool all your candy together and split it up into six equal shares. . . .

Trick or Treat! "Modeling" Equal Groupings

It's been a wildly successful Halloween night. The six of you counted up your loot, and you have a total of 852 candies—8 hundred-packs, 5 ten-packs, and 2 individual candies. If you divide them into 6 equal groups, how much will you each get? In other words, let's figure out 852 ÷ 6 = ? First, we'll set up 6 bowls to divide the candy into.

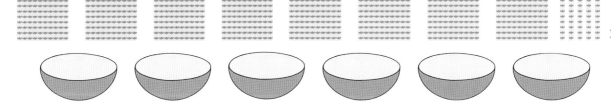

I mean, we could break apart ALL the hundred-packs and ten-packs and put the individual candies into the 6 bowls, one at a time, but that would take forever! Let's see if we can find a shorter way. Hmm, well, we could take 6 of the big hundred-packs and put **1 hundred** in each bowl, right? That would get us started, at least.

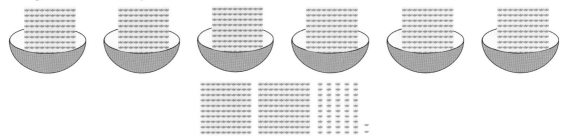

Gosh, the only way to split up those remaining 2 hundred-packs would be to do just that—split them up! So we'll break them into *20 little ten-packs*. And now we have a total of 25 little ten-packs, see?

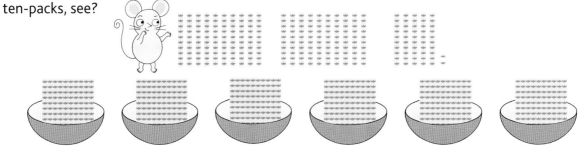

Well, we can't divide up the 25 ten-packs evenly to all 6 bowls, because 6 doesn't go into 25 evenly. But we can certainly put 24 of those ten-packs into the 6 bowls evenly: Since 24 ÷ 6 = **4**, we know each bowl can get **4 ten-packs** of candies, right?

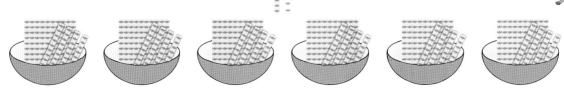

We've got one ten-pack left over, so we'll have to break up that remaining ten-pack into *10 individual candies*. Now all that's left to divide up is 12 individual candies!

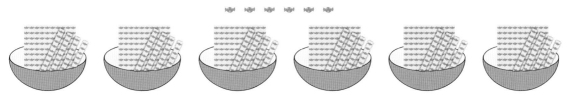

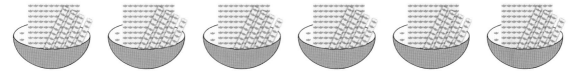

How can we split up the 12 individual candies into the 6 bowls? That's easy! Since 12 ÷ 6 = 2, that means each bowl will also get **2** candies, and we're done filling the bowls!

So, how much is in each bowl? Each bowl has **1** hundred-pack, **4** ten-packs, and **2** individual candies: **142**. So that means **852 ÷ 6 = 142**. Nice!

If the Halloween candy problem had 853 candies instead of 852 candies, we could do it the same exact way, but then at the *very* end, we'd be dividing up 13 candies instead of 12, right? After we put 2 in each bowl, we would be left with 1 extra candy (which wouldn't go in a bowl), and so our answer would have a remainder of 1, like this: **853 ÷ 6 = 142 R1.**

> OKAY, I LOVE THE CANDY ANGLE, BUT IF WE'RE GONNA DO PROBLEMS LIKE THIS, DO WE REALLY HAVE TO DRAW LIKE 800 CANDIES EACH TIME? THAT'S GONNA TAKE FOREVER.

> WE WILL BE DOING SOME DRAWING, BUT *PLACE VALUE* IS GOING TO HELP US SO WE DON'T HAVE TO DRAW AS MUCH. IT'S ANOTHER TYPE OF MODELING.

Big, Wide Bowls of Candy—
Modeling in a Place Value Chart

Let's try 531 ÷ 4 = ? This time, instead of drawing all the candies, we'll use a *place value chart* to keep track. Every "candy" we see in the hundreds column stands for an entire HUNDRED-pack of candies, every "candy" in the tens column stands for an entire TEN-pack of candies, and of course every candy in the ones column is just a single candy. We'll use cute little disks in the columns to be the candies (often called *place value disks*).

531 ÷ 4 = ?

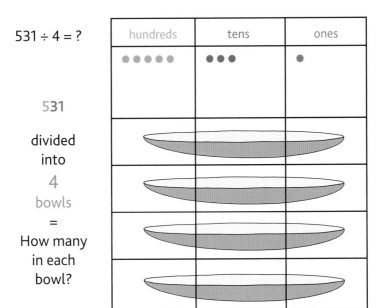

531

divided
into

4

bowls

=

How many
in each
bowl?

Let's start putting candy into bowls! We have 5 green hundred-packs of candies, so we can certainly put 1 hundred-pack into each of the **4 bowls**, right? Then we're left with 1 extra hundred—see what I mean? We'll break up that extra hundred-pack into *10 ten-packs*. It's a good idea to go back and cross off the disks we're done with and circle the disks that are moving:

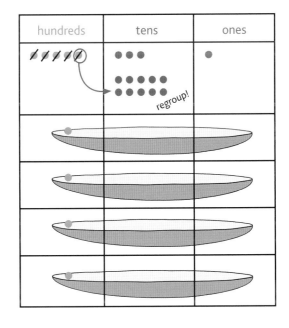

131 left to
divide up!

By the way, how much do we have left to divide up? We've divided up 400 into the bowls so far, which means we are left with 531 – 400 = 131 to go. (Yep, the 13 blue ten-packs *mean* 130 candies!) Make sure you understand that before moving on.

Okay, let's get back to dividing up this candy! Now that we're finished with the hundreds column, we'll move on to the tens column, where we see 13 disks. How do we divide 13 tens into 4 bowls? Well, we can definitely divide *12* evenly into 4 bowls, right? Since 12 ÷ 4 = **3**, we can put **3 ten-packs** into each bowl. Aren't they cute? And we'll be left with one ten-pack. So we'll break up that ten-pack and regroup it into *10 ones* (in the ones column), making sure to cross off all those ten-packs we have used or regrouped:

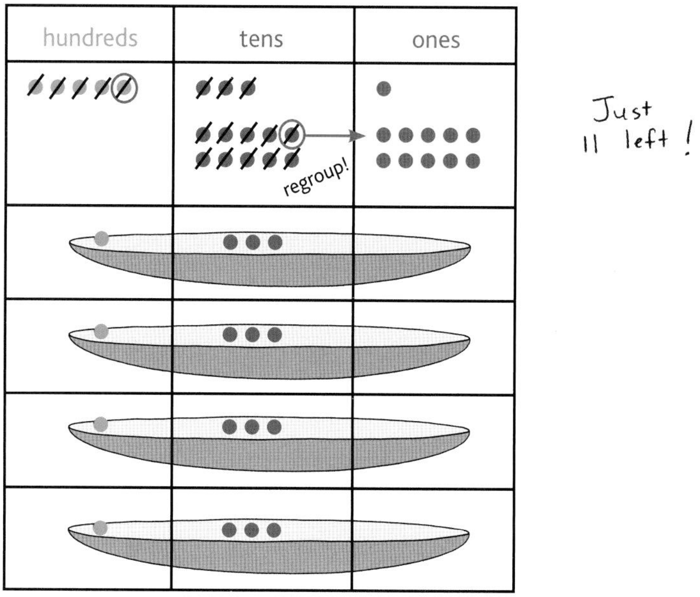

Finally, we're left with 11 individual candies to split up among the 4 bowls. (Gosh, if we had 12 candies, then we could put 3 in each bowl, but we don't—we only have 11.) Since 11 ÷ 4 = 2 **R3**, we can put **2 individual candies** into each bowl, with **3** candies remaining.

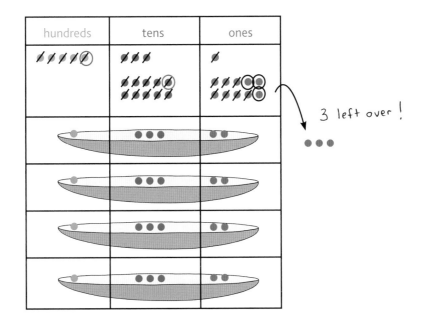

hundreds	tens	ones

3 left over!

So what's the answer? We were finding out how many candies go in *each bowl,* right? Looking at one of the bowls in our chart, that's 1 hundred-pack, 3 ten-packs, and 2 individual candies: 132! And we had **3** candies left over that couldn't be evenly divided into the 4 bowls— so that will be our remainder. In other words: **531 ÷ 4 = 132 R3**.

This place value "candy" chart method really comes down to splitting the candies into the bowls as evenly as we can, starting with the biggest chunks first, and working our way to the ones column. I mean, fair is fair, right?

When we regroup the disks, like when we rewrote 1 ten-pack as 10 individual candies: ⊘⟶ ⦙⦙⦙⦙⦙, notice that we drew them in the shape of a little rectangular *ten frame,* so it's easier to see that it's *ten* disks, which can save a lot of time and avoids confusion. See p. 23 in *Do Not Open This Math Book* for more on ten frames!

I HAVE A REALLY IMPORTANT QUESTION:
ARE THE BOWLS ALWAYS GOING TO BE PINK?
BECAUSE I REALLY LIKE PURPLE.

THEY CAN BE ANY COLOR YOU LIKE!

We drew bowls to make it easier to see the groups in this first example. But from this point, we'll just draw the horizontal *lines* to make separate groups for the disks to go in. Of course, we can always imagine our pink (or purple or any color!) bowls. We'll do one like that in #1 on p. 203. By the way, if you would like to see another example of place value disk division before you practice, take a sneak peek at p. 209!

Checking Our Answers!

Since multiplication and division are *inverse operations,* we can check our division answers with multiplication, and it even works with remainders! For example, to check **852 ÷ 6 = 142** from p. 197, we'd multiply 6 × 142 with whatever our favorite method is, and when we get **852**, we'd know we got it right!

Now let's see how this works when there's a remainder: To check **531 ÷ 4 = 132 R3** from p. 201, we would multiply 4 × 132 = 528, and then <u>after</u> the multiplication, we just *add the remainder:* 528 + 3 = **531**. Ta-da!

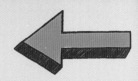

GAME TIME!

Do these division problems using the place value "candy" chart. In other words, model them with place value disks. I'll do the first one for you!

1. $209 \div 3 = \underline{?}$

Let's Play: First we'll draw our place value chart and write in the disks to show 209. That means 2 hundred-disks and 9 "ones" disks. And we'll draw lines underneath to make 3 empty rows for our groups that we'll be putting disks into. We can even imagine our short, wide pink bowls there (see p. 199). Let's divide: So, starting with the hundreds, how can we divide up 2 hundred-disks into 3 groups? We can't, because 2 is smaller than 3! So we'll have to regroup them into *20 ten-disks*.

	hundreds	tens	ones
209	• •		••••• ••••

3 groups (bowls!)

→

	hundreds	tens	ones
	⌀ ⌀ regroup!	•••••• •••••• •••••• ••	••••• ••••

That's better! So, how many of the 20 ten-disks can we split evenly among the 3 groups? Well, that's the same as asking *20 ÷ 3 = ?* See what I mean?

Now, when you turn the page, you're going to see a lot of dots. Don't worry if those disks look confusing—we're doing this *together*, remember? They're just little dots, and we're not going to let them scare us. . . .

Keep going! ⟶

Since 20 ÷ 3 = 6 R2, we can put 6 ten-disks into each group (see the left-hand chart below), making sure to keep them in the tens column. And we have 2 leftover blue ten-disks, right? So we'll regroup them as 20 red "ones" disks.

Make sure you see where that regrouping happens on the left-hand chart below.

hundreds	tens	ones
∅∅	////// ////// ////// regroup!	••••• •••• •••• •••• ••••

Our 3 groups! So far, they each have 6 tens in them.

	tens	ones
	••••• •	
	••••• •	
	••••• •	

→

hundreds	tens	ones
∅∅	////// ////// //////	///// //(∅∅) ///// /////

Two left over!

Now each of our 3 groups has 6 tens and 9 ones in them!

	tens	ones
	••••• •	•••• ••••
	••••• •	•••• ••••
	••••• •	•••• ••••

Now, ONLY looking above at the chart on the left, we ask: How many of the <u>29</u> red disks can we split up evenly among our 3 groups? Since 29 ÷ 3 = 9 **R2**, we can put 9 disks in each group (see the chart above on the right), with 2 leftover disks, which becomes our remainder. So, how many ended up in *each* group? 69, with **2** left over. Done!

Answer: 209 ÷ 3 = 69 R2

 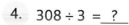

2. 369 ÷ 3 = __?__

3. 848 ÷ 4 = __?__

4. 308 ÷ 3 = __?__

5. 255 ÷ 5 = __?__

6. 531 ÷ 4 = __?__

7. 624 ÷ 6 = __?__

(Answers on page 223.)

High-Speed Freeway Lanes:
The Traditional Method, aka Long Division

Yep, just like with multiplication, after we do all the pretty pictures and models, there's a "traditional" way that involves less writing and more thinking—and is much faster once you get the hang of it.

Remember our dividing houses from p. 61? We're going to bring them back! Let's say we want to divide 465 ÷ 5 = ? We'll write it as a dividing house, and I like to write freeway "lanes" to keep all the place values straight:

Just like in our other methods, we divide by looking at the *biggest chunk first*—that's the hundreds column. We ask, "Does 5 go into **4**?" Nope! So we don't write anything in the hundreds column. Notice that if we were using place value disks, we would now break up the 4 hundreds into 40 tens, to get a total of 46 *tens*!

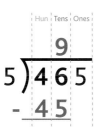

Here, we do something similar! We start *thinking* about the 4 as *40* now, and we ask: Can 5 divide into **46**? Sure! 5 goes into 46 a bunch of times. How many times? Well . . . what's the biggest multiple of 5 that *isn't bigger than* 46? We know that 5 × **9** = 45, so the 5 fits into 46 at least **9** times! Next, we'll check 5 × 10 = 50—oops, too big (the whipped cream fell off!) so we know that **9** is the right number of 5's. We write the 9 above, in the tens column.

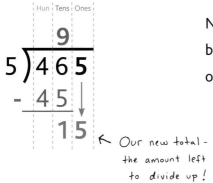

← *Our new total — the amount left to divide up!*

Next, we subtract the 45 from the 46 and get **1**, and then bring down the ones digit **5** to join it, to get **15**. And that's our "new total," the stuff still left to divide up!

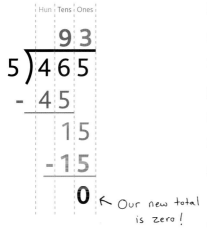

← *Our new total is zero!*

Now we start again, with our new total left to be divided up: 15. Does 5 go into 15? Yes! We know from our multiplication facts that 5 × 3 = 15, so that means 5 goes into 15 exactly **3** times. We write that 3 in the ones column, and then subtract, like we did before, to see how many we still have left to divide up. And since there are *no more numbers to bring down*—we're done!

Answer: 465 ÷ 5 = 93

In long division, if we ever subtract and get a number bigger than the divisor, that means *we didn't pick a big enough number* to go on top! For example, if we're dividing 652 ÷ 8, after realizing 8 was too big to go into 6, we'd ask, "How many times does 8 go into 65?" If we worked too quickly and said, "Oh, 8 × 7 is 56, which isn't too much smaller than 65," as soon as we did our subtraction 65 − 56 = 9, we'd realize 9 is bigger than 8. Oops! That means 8 goes into 65 *more* than 7 times—in fact, you might notice that 8 can fit into 65 a total of 8 times, because 8 × 8 = 64!

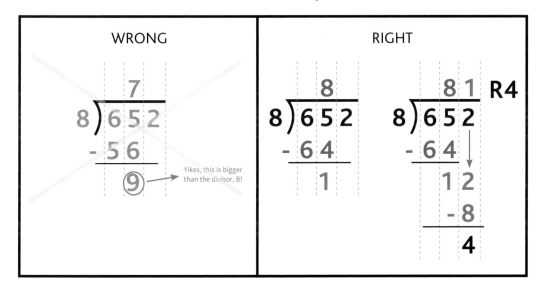

Always think about our multiplication facts (and whipped cream—see p. 159!), then check a few before deciding which one will work best. You'll do great!

Candy and Freeways: A Side-by-Side

Now we'll look at this traditional method side by side with the place value "candy" method!

Long Division: 561 ÷ 3

So 561 is our dividend (inside the house) and 3 is the divisor. We can also use dotted lines to keep our place value columns straight!

$$3\overline{)561}$$

How many times does 3 go into 5?   1 time!

We write the 1 up top, in the hundreds column. Then we subtract 5 – 3 = 2 in the hundreds column, and "bring down" the 6 to get 26.

$$
\begin{array}{r}
1 \\
3\overline{)561} \\
-3\downarrow \\
\hline
26
\end{array}
$$

KNOWING OUR MULTIPLICATION FACTS REALLY COMES IN HANDY!

Next: How many times does 3 go into 26?

8 times!
(We know this since 3 × 8 = 24, and 3 × 9 = 27 is too big—the whipped cream would fall off!)

We write the 8 up top, in the tens column. Then we subtract 26 – 24 = 2, and "bring down" the 1 to get 21.

$$
\begin{array}{r}
18 \\
3\overline{)561} \\
-3\downarrow \\
\hline
26 \\
-24\downarrow \\
\hline
21
\end{array}
$$

SO "BRINGING DOWN" NUMBERS HELPS US SEE HOW MUCH IS LEFT TO DIVIDE.

How many times does 3 go into 21?
Exactly 7 times!
(We know this since 3 × 7 = 21.)

We write the 7 up top, in the ones column. Then we subtract 21 – 21. There's nothing left to "bring down," so we're done dividing, and we got zero, so there's no remainder.

561 ÷ 3 = 187

Done!

$$
\begin{array}{r}
187 \\
3\overline{)561} \\
-3\downarrow \\
\hline
26 \\
-24\downarrow \\
\hline
21 \\
-21 \\
\hline
21 \\
\hline
0
\end{array}
$$

Comparison of Long Division

It'll be fun! Let's do 561 ÷ 3 both ways!

Place Value "Candy" Chart: 561 ÷ 3

First, we set up our place value chart, with 561 written as disks and 3 bowls to divide it up into!

How many of the 5 hundred-disks can go into 3 bowls?   1 each!

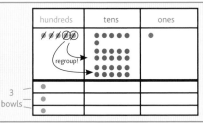

We write 1 hundred-disk in the hundreds column of each "bowl." Then we cross off the 3 hundreds we moved, and regroup the other 2 into 20 tens, for a total of 26 tens.

Next: How many of the 26 ten-disks can go into each of the 3 bowls ?

8 each!

(We know this since 26 ÷ 3 = 8 R2. And the 2 leftover ten-disks will get regrouped next!)

> IN BOTH METHODS, ONLY 24 OF THE 26 CAN BE DIVIDED UP INTO 3 EQUAL PARTS!

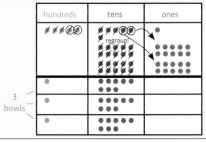

We write 8 ten-disks in the tens column of each "bowl." Then we cross off the 24 tens we moved, and regroup the leftover 2 into 20 ones, for a total of 21 ones.

> IN BOTH METHODS, THE BIG NUMBER EVENTUALLY GETS DIVIDED UP INTO 3 EQUAL AMOUNTS.

How many of the 21 ones can go into the 3 bowls?
Exactly 7 each!
(We know this since 3 × 7 = 21.)

We write 7 "ones" disks in the ones column of each "bowl." Then we cross off the 21 ones we moved. Nothing's left to divide up, and there was no remainder, so we're done! In *each bowl*, there's 1 hundred, 8 tens, and 7 ones, so:

$$561 ÷ 3 = 187$$

Done!

STEP BY STEP:

On the Freeway—a Step by Step of Long Division

Now that you've seen the long division "freeway" method and how it compares to our place value "candy" method, here's a little Step by Step of long division to help make it extra clear, but don't worry if you need to read it a few times before it sinks in. Just remember, there are four steps in long division that repeat: Divide! Multiply! Subtract! Bring down! We'll do **459 ÷ 7 = ?**

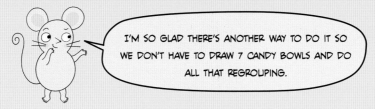

I'M SO GLAD THERE'S ANOTHER WAY TO DO IT SO WE DON'T HAVE TO DRAW 7 CANDY BOWLS AND DO ALL THAT REGROUPING.

Step 1. DIVIDE! How many times does the divisor go into the first digit of the dividend, or if it doesn't, how many times does the divisor go into the first *two* digits?

Write that number on top, directly over the *last* digit of the number you divided the divisor into!

We can't do 4 ÷ 7, but we can do 45 ÷ 7! Checking our multiplication facts, 7 × 6 = 42 works, but 7 × 7 = 49 is too big. We'll write the 6 directly above the 5— the *last* digit in 45!

$$7 \overline{)459} \quad 6$$

Step 2. MULTIPLY! Now multiply that number on top times the divisor, and write it below.

Here, we'll do: 7 × 6 = 42

$$7 \overline{)459} \quad \begin{matrix} 6 \\ 42 \end{matrix}$$

Step 3. SUBTRACT! Subtract the two numbers.

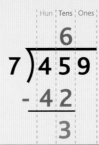

IT MAKES SENSE THAT WE'D SUBTRACT OFF THE BIG CHUNK WE ALREADY DIVIDED UP--THE 42--BECAUSE WE DON'T NEED TO DIVIDE THAT PART UP ANYMORE!

"42" is how much we've divided up already (of course it's really 420), so we'll subtract that off and see what's left:

45 – 42 = 3 (which is really 30!)

$$
\begin{array}{r}
\text{Hun}\ |\ \text{Tens}\ |\ \text{Ones} \\
6 \\
7\,)\overline{4\ 5\ 9} \\
-\ 4\ 2 \\
\hline
3
\end{array}
$$

Step 4. BRING DOWN! Use an arrow to "bring down" the next digit of the dividend. That's our "new total" we'll use to divide into next. (We'll never need to use the full dividend again.)

$$
\begin{array}{r}
\text{Hun}\ |\ \text{Tens}\ |\ \text{Ones} \\
6 \\
7\,)\overline{4\ 5\ 9} \\
-\ 4\ 2\ \downarrow \\
\hline
3\ 9
\end{array}
$$

Our "new total"!

We're done with the full "459"—we'll just use the 39 now.

Step 5. REPEAT! Then we repeat the above four steps *until there are no more numbers to bring down*. If our last subtraction problem doesn't give us zero, then that's our remainder. Done!

DIVIDE! MULTIPLY! SUBTRACT! BRING DOWN! DIVIDE! MULTIPLY! SUBTRACT! BRING DOWN!

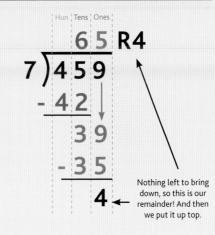

$$
\begin{array}{r}
\text{Hun}\ |\ \text{Tens}\ |\ \text{Ones} \\
6\ 5\ \text{R4} \\
7\,)\overline{4\ 5\ 9} \\
-\ 4\ 2\ \downarrow \\
\hline
3\ 9 \\
-\ 3\ 5 \\
\hline
4
\end{array}
$$

Nothing left to bring down, so this is our remainder! And then we put it up top.

Answer: 459 ÷ 7 = 65 R4

Arrows vs. "Zeros Show Below"

Instead of drawing arrows, we can fill in <u>zeros</u> for the subtraction part of long division, which is a more accurate description of what's going on! For example, on p. 208, when we subtracted 3 from 5, we were in the *hundreds* column, so we were *really* subtracting 300, right? Well, if we write out the full 300, then we don't even need arrows. Take a look:

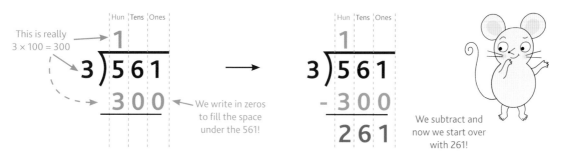

This is really
3 × 100 = 300

We write in zeros
to fill the space
under the 561!

We subtract and
now we start over
with 261!

Now we treat the 261 like the beginning of the problem, and we say "How many times does 3 go into 2? None. How about 26?" and then since 3 × **8** = 24, we write **8** above, and subtract 24, filling in the zero. Then we get:

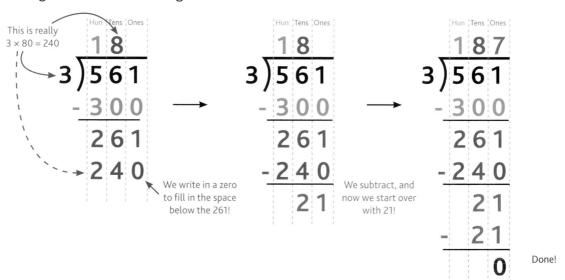

This is really
3 × 80 = 240

We write in a zero
to fill in the space
below the 261!

We subtract, and
now we start over
with 21!

Done!

And of course, we get the same answer as on p. 208. If you like this more than the arrows, try using it for the Game Time on the next page! With this "zeros show below" method, the four long division steps become: **Divide! Multiply! Fill in zeros! Subtract!**

GAME TIME!

Divide, using the traditional/long division method steps on p. 210.
Or feel free to use the "zeros show below" method instead (see p. 212),
and use freeway lines if you want. I'll do the first one for you!

1. $4\overline{)368}^{?}$

Let's Play: Looking at the steps on p. 210: **Step 1.** Divide! So, does 4 go into 3? Nope! Then does 4 go into 36? Yep! Since 4 × 9 = 36, we'll write 9 directly above the 6, the last digit in 36. **Step 2.** Multiply! We multiply the number on top times the divisor: 9 × 4 = **36,** and write that **36** below the 36 in the dividend. **Step 3.** Subtract! We subtract 36 – 36 = **0.** But just because we got zero doesn't mean we're done—we still have to bring down that 8! **Step 4.** Bring it down! We bring down the 8, and now our new total is "**08**," in other words: **8**!

$$4\overline{)368}\atop{-36\atop{0}}$$ Wait! Not done yet! $$4\overline{)368}\atop{-36\atop{08}}$$ $$4\overline{)368}^{92}\atop{-36\atop{08\atop{-8\atop{0}}}}$$

Now we **REPEAT** and start over with **Step 1.** Divide! So, does 4 go into 8? Yes! It goes in exactly 2 times, so we write **2** above. **Step 2.** Multiply! We multiply 4 × 2 = 8 and write that 8 below. **Step 3.** Subtract! We subtract 8 – 8 = 0. Since there's nothing left to bring down, we're done dividing. And since our last subtraction gave us zero, there's no remainder, and we get 92. Done!

Keep going! ⟶

Answer: 368 ÷ 4 = 92

213

By the way, here's how it would look with the "zeros show below" method from p. 212—it actually ends up being shorter:

This is really
4 × 90 = 360

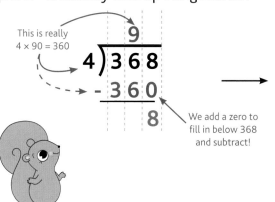

We add a zero to fill in below 368 and subtract!

$$4\overline{)368}$$

9
−360
8

92
4)368
−360
8
− 8
0

Done!

Use either method for these!

2. $3\overline{)69}$?

3. $4\overline{)88}$?

4. $4\overline{)432}$?

5. $5\overline{)555}$?

6. $6\overline{)567}$?

7. $5\overline{)875}$?

8. $5\overline{)788}$?

9. $2\overline{)98}$?

10. $3\overline{)405}$?

11. $772 \div 9 = $?

Hint: Write these with dividing houses first!

12. $874 \div 2 = $?

13. $999 \div 3 = $?

14. $749 \div 7 = $?

15. $391 \div 8 = $?

16. $313 \div 3 = $?

17. $4,367 \div 4 = $?

Hint: These work the same way—you got this!

18. $3,910 \div 8 = $?

19. $4,599 \div 7 = $?

(Answers on page 223.)

Most textbooks don't put "freeway lines" on division, but it's a really helpful way to keep everything straight. I bet your teacher will love it!

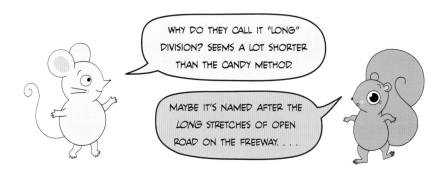

WHY DO THEY CALL IT "LONG" DIVISION? SEEMS A LOT SHORTER THAN THE CANDY METHOD.

MAYBE IT'S NAMED AFTER THE *LONG* STRETCHES OF OPEN ROAD ON THE FREEWAY. . . .

For more long division methods, like array/area models, equal groupings, and the Big 7, check out **TheTimesMachine.com**. And for my Most Fabulous Review of Long Division Ever and to see more examples of the traditional long division method, including with two-digit divisors, visit **mathdoesntsuck.com/extras**! In fact, once you've mastered the concepts in this book, you'll be ready for my next book, *Math Doesn't Suck,* for things like fractions, decimals, and more!

My Eternal Gratitude

Thank you to my parents, Mahaila and Chris, for always encouraging and believing in me! Thank you to my best friend in the whole world, Crystal, and to her husband, Mike, and their darling children, Cricket and little Cayden! Thank you to the rest of my wonderful family, including but not limited to Chris Jr., Connor, Lorna (and kids!), Molly, and the Svesloskys, Sims, and Vertas. And to Grammy (1910–2010) and Opa (1919–2016), who were always so proud of me, and whose voicemails are still saved on my phone to prove it. Thank you, Cheryl, for all your help to our family in so many ways, and thank you, Kim, for just everything! An enormous thank-you to Mike Verta for being a great co-parenting partner, and for another phenomenal title and cover.

Thank you to my wonderful illustrator, Josée Masse, for the adorable and often very complicated comics. Thank you to everyone at PRH, including Phoebe Yeh and the incomparable Barbara Marcus, for your vision and belief in me. Thank you to my talented (and tireless!) editor, Emily Easton. I'm so grateful for your partnership on all of our books together! Thank you to Elizabeth Tardiff, Nicole de las Heras, and Monique Razzouk for all the involved design work in this book, and Alison Kolani, Megan Williams, Claire Nist, Amy Bowman, and more, for bringing it all together. All your hard work is so very appreciated. These books are more complex than most realize, and it takes a village.

Thank you to my longtime literary agent, Laura Nolan, for first seeking me out 15 years ago to start writing books to inspire kids in math, and for encouraging me to continue with a younger crowd. Thank you to my literary lawyer, Stephen Breimer, and also to Cathey Lizzio,

Pat Brady, and Matt Sherman, for your many years of support, and thank you to Noreen Herits for helping me get the word out about my books. Thank you to all my friends at Hallmark Channel for working my movies around my crazy book schedule! Thank you to the educators who read my early drafts many moons ago and gave feedback and insight into today's classrooms, including Laura Chambless, Wendy Lucka, Betsy Schuman, Jen Stern, Moira Talan, Matt Persek, Alina Marsh, Mariko Garcia, and Tracey Evans. Thank you to Connor McKellar for your help. Thank you to everyone who proofread this book, including my mom, Mahaila McKellar; Lisa Swerdlow; Tracey Evans; Mike Scafati; Kayo Goto; and last but not least, my longtime friend and master proofreader from whom no error escapes, Jonathan Farley.

Thank you to my amazing, loving husband, Scott—I love you so much and feel so lucky to be married to you. Thank you to Hunter for being a wonderful stepson (and awesome football player), and for even letting me help you with math sometimes. Finally, thank you to my precious boy, Draco. You have changed me forever, in the best ways possible, and it is my greatest honor to watch you grow and help you thrive. I love you!

Answer Key

Chapter 1

p. 20   2. 2 rows, 8 columns, 2 x 8 = 16   3. 3 rows, 3 columns, 3 x 3 = 9   4. 3 rows, 4 columns, 3 x 4 = 12   5. 4 rows, 3 columns, 4 x 3 = 12
6. 3 rows, 5 columns, 3 x 5 = 15   7. 4 rows, 6 columns, 4 x 6 = 24   8. 5 rows, 5 columns, 5 x 5 = 25   9. 3 rows, 7 columns, 3 x 7 = 21
10. 4 rows, 9 columns, 4 x 9 = 36

p. 25   2. factors: 4 and 6, product: 24, 6 x 4 = 24   3. factors: 5 and 3, product: 15, 3 x 5 = 15   4. factors: 4 and 6, product: 24, 24 = 6 x 4
5. factors: 6 and 7, product: 42, 7 x 6 = 42   6. factors: 7 and 8, product: 56, 56 = 8 x 7   7. factors: 0 and 7, product: 0, 7 x 0 = 0
8. factors: 10 and 2, product: 20, 20 = 2 x 10   9. factors: 8 and 6, product: 48, 6 x 8 = 48   10. factors: 7 and 9, product: 63, 9 x 7 = 63
11. factors: 3 and 7, product: 21, 21 = 7 x 3   12. factors: 5 and 6, product: 30, 30 = 6 x 5   13. factors: 1 and 3, product: 3, 3 x 1 = 3

p. 27   2. A   3. D   4. B   5. E

Chapter 2

p. 37   2. 5 x 4 = 20   3. 3 x 7 = 21   4. 6 x 3 = 18   5. 5 x 9 = 45   6. 4 x 8 = 32   7. 3 x 5 = 15   8. 9 x 9 = 81   9. 10 x 2 = 20   10. 6 x 6 = 36
11. 9 x 3 = 27   12. 6 x 7 = 42   13. 8 x 6 = 48   14. 7 x 7 = 49   15. 5 x 1 = 5   16. 8 x 7 = 56   17. 12 x 4 = 48   18. 7 x 4 = 28   19. 8 x 4 = 32
20. 8 x 8 = 64   21. 10 x 9 = 90   22. 4 x 6 = 24   23. 7 x 5 = 35   24. 11 x 6 = 66   25. 7 x 8 = 56   26. 9 x 7 = 63   27. 12 x 12 = 144
28. 11 x 8 = 88   29. 11 x 12 = 132

Chapter 3

p. 45   2. 7 x 13 = (7 x 3) + (7 x 10) = 91   3. 5 x 15 = (5 x 5) + (5 x 10) = 75   4. 6 x 14 = (6 x 10) + (6 x 4) = 84   5. 4 x 16 = (4 x 10) + (4 x 6) = 64
6. 4 x 17 = (4 x 7) + (4 x 10) = 68   7. 8 x 12 = (8 x 2) + (8 x 10) = 96

Chapter 4

p. 50   2. 12 ÷ 4 = 3   3. 15 ÷ 3 = 5   4. 12 ÷ 3 = 4   5. 10 ÷ 5 = 2   6. 16 ÷ 4 = 4   7. 16 ÷ 2 = 8   8. 12 ÷ 4 = 3   9. 21 ÷ 3 = 7
10. 15 ÷ 3 = 5 (or 15 ÷ 5 = 3)

p. 56   2. 2 x 3 = 6, 3 x 2 = 6, 6 ÷ 2 = 3, 6 ÷ 3 = 2   3. 10 x 9 = 90, 9 x 10 = 90, 90 ÷ 10 = 9, 90 ÷ 9 = 10   4. 7 x 8 = 56, 8 x 7 = 56, 56 ÷ 7 = 8, 56 ÷ 8 = 7
5. 6 x 5 = 30, 5 x 6 = 30, 30 ÷ 6 = 5, 30 ÷ 5 = 6   6. 9 x 2 = 18, 2 x 9 = 18, 18 ÷ 9 = 2, 18 ÷ 2 = 9   7. 6 x 7 = 42, 7 x 6 = 42, 42 ÷ 6 = 7, 42 ÷ 7 = 6
8. 4 x 1 = 4, 1 x 4 = 4, 4 ÷ 4 = 1, 4 ÷ 1 = 4   9. 8 x 6 = 48, 6 x 8 = 48, 48 ÷ 8 = 6, 48 ÷ 6 = 8   10. 5 x 11 = 55, 11 x 5 = 55, 55 ÷ 5 = 11, 55 ÷ 11 = 5
11. 8 x 4 = 32, 4 x 8 = 32, 32 ÷ 8 = 4, 32 ÷ 4 = 8   12. 6 x 6 = 36, 36 ÷ 6 = 6   13. 63 ÷ 7 = 9, 63 ÷ 9 = 7, 7 x 9 = 63, 9 x 7 = 63
14. 28 ÷ 7 = 4, 28 ÷ 4 = 7, 7 x 4 = 28, 4 x 7 = 28   15. 84 ÷ 7 = 12, 84 ÷ 12 = 7, 7 x 12 = 84, 12 x 7 = 84
16. 54 ÷ 9 = 6, 54 ÷ 6 = 9, 9 x 6 = 54, 6 x 9 = 54   17. 18 ÷ 6 = 3, 18 ÷ 3 = 6, 6 x 3 = 18, 3 x 6 = 18
18. 132 ÷ 11 = 12, 132 ÷ 12 = 11, 11 x 12 = 132, 12 x 11 = 132   19. 27 ÷ 3 = 9, 27 ÷ 9 = 3, 3 x 9 = 27, 9 x 3 = 27
20. 100 ÷ 10 = 10, 10 x 10 = 100   21. 12 x 12 = 144, 144 ÷ 12 = 12

p. 59   2. 7 x ? = 49 and 49 ÷ 7 = 7   3. 2 x ? = 12 and 12 ÷ 2 = 6   4. 6 x ? = 12 and 12 ÷ 6 = 2   5. 5 x ? = 20 and 20 ÷ 5 = 4
6. 4 x ? = 20 and 20 ÷ 4 = 5   7. 4 x ? = 24 and 24 ÷ 4 = 6   8. 6 x ? = 24 and 24 ÷ 6 = 4   9. 7 x ? = 63 and 63 ÷ 7 = 9
10. 10 x ? = 110 and 110 ÷ 10 = 11   11. 11 x ? = 110 and 110 ÷ 11 = 10   12. 8 x ? = 56 and 56 ÷ 8 = 7   13. 7 x ? = 56 and 56 ÷ 7 = 8

p. 64   2. 3   3. 2   4. 10   5. 10   6. 8   7. 7   8. 4   9. 11   10. 7   11. 8   12. 9   13. 6
14. 6   15. 5   16. 12   17. 9   18. 6   19. 9   20. 5   21. 12   22. 11   23. 8   24. 9   25. 11

p. 77   2. 2 x 2 = 4   3. 5 x 0 = 0   4. 1 x 6 = 6   5. 5 x 2 = 10   6. 11 x 2 = 22   7. 8 x 1 = 8   8. 10 x 2 = 20   9. 2 x 8 = 16
10. 2 x 3 = 6   11. 6 x 2 = 12   12. 0 x 1 = 0   13. 8 x 2 = 16   14. 2 x 4 = 8   15. 8 x 0 = 0   16. 2 x 9 = 18   17. 7 x 2 = 14
18. 1 x 1 = 1   19. 12 x 1 = 12   20. 2 x 12 = 24   21. 0 x 0 = 0

p. 78   2. 2 ÷ 1 = 2   3. 0 ÷ 5 = 0   4. 6 ÷ 6 = 1   5. 10 ÷ 2 = 5   6. 22 ÷ 11 = 2   7. 8 ÷ 1 = 8   8. 20 ÷ 2 = 10   9. 16 ÷ 2 = 8
10. 6 ÷ 3 = 2   11. 12 ÷ 2 = 6   12. 0 ÷ 1 = 0   13. 16 ÷ 8 = 2   14. 8 ÷ 2 = 4   15. 12 ÷ 1 = 12   16. 18 ÷ 2 = 9   17. 14 ÷ 7 = 2
18. 1 ÷ 1 = 1   19. 12 ÷ 12 = 1   20. 24 ÷ 12 = 2   21. 18 ÷ 9 = 2

p. 85   2. 3 x 3 = 9   3. 3 x 6 = 18   4. 3 x 4 = 12   5. 3 x 8 = 24   6. 3 x 7 = 21   7. 4 x 3 = 12   8. 5 x 3 = 15   9. 3 x 0 = 0
10. 3 x 2 = 6   11. 8 x 3 = 24   12. 3 x 5 = 15   13. 7 x 3 = 21   14. 2 x 3 = 6   15. 5 x 3 = 15   16. 1 x 3 = 3   17. 6 x 3 = 18

p. 86   2. 6 ÷ 2 = 3   3. 15 ÷ 3 = 5   4. 18 ÷ 3 = 6   5. 6 ÷ 3 = 2   6. 15 ÷ 5 = 3   7. 0 ÷ 3 = 0   8. 21 ÷ 7 = 3   9. 24 ÷ 3 = 8
10. 9 ÷ 3 = 3   11. 3 ÷ 1 = 3   12. 21 ÷ 3 = 7   13. 12 ÷ 4 = 3   14. 24 ÷ 8 = 3   15. 12 ÷ 3 = 4   16. 18 ÷ 6 = 3

p. 90   2. 4 x 4 = 16   3. 4 x 7 = 28   4. 4 x 6 = 24   5. 4 x 8 = 32   6. 4 x 3 = 12   7. 2 x 4 = 8   8. 1 x 4 = 4   9. 4 x 0 = 0
10. 5 x 4 = 20   11. 4 x 1 = 4   12. 4 x 5 = 20   13. 3 x 4 = 12   14. 8 x 4 = 32   15. 4 x 2 = 8   16. 6 x 4 = 24   17. 7 x 4 = 28

p. 91   2. 16 ÷ 4 = 4   3. 20 ÷ 4 = 5   4. 24 ÷ 6 = 4   5. 4 ÷ 2 = 2   6. 12 ÷ 3 = 4   7. 4 ÷ 4 = 1   8. 0 ÷ 4 = 0   9. 32 ÷ 4 = 8
10. 12 ÷ 4 = 3   11. 20 ÷ 5 = 4   12. 28 ÷ 7 = 4   13. 24 ÷ 4 = 6   14. 28 ÷ 4 = 7   15. 4 ÷ 1 = 4   16. 8 ÷ 2 = 4

p. 96   2. 5 x 4 = 20   3. 1 x 5 = 5   4. 5 x 6 = 30   5. 5 x 3 = 15   6. 7 x 5 = 35   7. 0 x 5 = 0   8. 8 x 5 = 40   9. 5 x 5 = 25
10. 2 x 5 = 10   11. 6 x 5 = 30   12. 3 x 5 = 15   13. 4 x 5 = 20   14. 5 x 8 = 40   15. 5 x 2 = 10   16. 5 x 7 = 35   17. 5 x 0 = 0

p. 97   2. 50 ÷ 5 = 10   3. 25 ÷ 5 = 5   4. 30 ÷ 5 = 6   5. 10 ÷ 5 = 2   6. 15 ÷ 5 = 3   7. 40 ÷ 8 = 5   8. 20 ÷ 4 = 5   9. 40 ÷ 5 = 8
10. 15 ÷ 3 = 5   11. 30 ÷ 6 = 5   12. 35 ÷ 5 = 7   13. 0 ÷ 5 = 0   14. 35 ÷ 7 = 5   15. 5 ÷ 1 = 5   16. 45 ÷ 5 = 9   17. 10 ÷ 2 = 5

p. 100   2. 6 x 4 = 24   3. 6 x 8 = 48   4. 6 x 3 = 18   5. 6 x 7 = 42   6. 6 x 6 = 36   7. 3 x 6 = 18   8. 1 x 6 = 6   9. 6 x 2 = 12
10. 6 x 0 = 0   11. 7 x 6 = 42   12. 6 x 5 = 30   13. 8 x 6 = 48   14. 6 x 1 = 6   15. 2 x 6 = 12   16. 5 x 6 = 30   17. 4 x 6 = 24

p. 101   2. 24 ÷ 4 = 6   3. 30 ÷ 6 = 5   4. 24 ÷ 6 = 4   5. 12 ÷ 2 = 6   6. 18 ÷ 3 = 6   7. 48 ÷ 8 = 6   8. 6 ÷ 6 = 1   9. 48 ÷ 6 = 8
10. 18 ÷ 6 = 3   11. 0 ÷ 6 = 0   12. 42 ÷ 6 = 7   13. 6 ÷ 1 = 6   14. 42 ÷ 7 = 6   15. 48 ÷ 6 = 8   16. 12 ÷ 6 = 2

p. 104   2. 7 x 6 = 42   3. 7 x 3 = 21   4. 7 x 4 = 28   5. 7 x 8 = 56   6. 7 x 7 = 49   7. 0 x 7 = 0   8. 1 x 7 = 7   9. 6 x 7 = 42
10. 8 x 7 = 56   11. 7 x 2 = 14   12. 7 x 5 = 35   13. 4 x 7 = 28   14. 3 x 7 = 21   15. 2 x 7 = 14   16. 5 x 7 = 35   17. 7 x 0 = 0

p. 105   2. 28 ÷ 4 = 7   3. 35 ÷ 7 = 5   4. 28 ÷ 7 = 4   5. 14 ÷ 2 = 7   6. 21 ÷ 3 = 7   7. 63 ÷ 9 = 7   8. 70 ÷ 7 = 10   9. 56 ÷ 8 = 7
10. 21 ÷ 7 = 3   11. 7 ÷ 7 = 1   12. 42 ÷ 7 = 6   13. 0 ÷ 7 = 0   14. 49 ÷ 7 = 7   15. 63 ÷ 9 = 7   16. 56 ÷ 7 = 8

p. 107   2. 8 x 4 = 32   3. 8 x 3 = 24   4. 8 x 6 = 48   5. 8 x 8 = 64   6. 8 x 7 = 56   7. 1 x 8 = 8   8. 3 x 8 = 24   9. 7 x 8 = 56
10. 5 x 8 = 40   11. 8 x 2 = 16   12. 8 x 5 = 40   13. 0 x 8 = 0   14. 4 x 8 = 32   15. 2 x 8 = 16   16. 6 x 8 = 48   17. 8 x 1 = 8

p. 108   2. 32 ÷ 4 = 8   3. 40 ÷ 8 = 5   4. 32 ÷ 8 = 4   5. 16 ÷ 2 = 8   6. 48 ÷ 6 = 8   7. 56 ÷ 7 = 8   8. 24 ÷ 3 = 8   9. 56 ÷ 8 = 7
10. 24 ÷ 8 = 3   11. 0 ÷ 8 = 0   12. 48 ÷ 8 = 6   13. 8 ÷ 8 = 1   14. 64 ÷ 8 = 8   15. 40 ÷ 5 = 8   16. 16 ÷ 8 = 2   17. 8 ÷ 1 = 8

p. 115   2. 3 x 9 = 27   3. 1 x 9 = 9   4. 6 x 9 = 54   5. 7 x 9 = 63   6. 9 x 2 = 18   7. 9 x 5 = 45   8. 4 x 9 = 36   9. 8 x 9 = 72
10. 2 x 9 = 18   11. 9 x 9 = 81   12. 5 x 9 = 45   13. 9 x 3 = 27   14. 9 x 8 = 72   15. 9 x 0 = 0   16. 9 x 7 = 63   17. 9 x 6 = 54

p. 116   2. 36 ÷ 4 = 9   3. 45 ÷ 9 = 5   4. 36 ÷ 9 = 4   5. 18 ÷ 2 = 9   6. 72 ÷ 8 = 9   7. 63 ÷ 7 = 9   8. 72 ÷ 9 = 8   9. 63 ÷ 9 = 7
10. 0 ÷ 9 = 0   11. 27 ÷ 9 = 3   12. 54 ÷ 9 = 6   13. 9 ÷ 9 = 1   14. 81 ÷ 9 = 9   15. 45 ÷ 5 = 9   16. 27 ÷ 3 = 9   17. 54 ÷ 6 = 9

p. 121   2. 3 x 10 = 30   3. 1 x 10 = 10   4. 6 x 10 = 60   5. 11 x 10 = 110   6. 10 x 2 = 20   7. 10 x 5 = 50   8. 12 x 10 = 120   9. 8 x 10 = 80   10. 25 x 10 = 250   11. 17 x 10 = 170   12. 82 x 10 = 820   13. 10 x 54 = 540   14. 10 x 88 = 880   15. 10 x 0 = 0   16. 10 x 23 = 230   17. 53 x 10 = 530   18. 44 x 10 = 440   19. 10 x 12 = 120   20. 10 x 72 = 720   21. 10 x 30 = 300   22. 77 x 10 = 770

p. 122   2. 80 ÷ 10 = 8   3. 50 ÷ 10 = 5   4. 100 ÷ 10 = 10   5. 120 ÷ 10 = 12   6. 110 ÷ 10 = 11   7. 630 ÷ 10 = 63   8. 900 ÷ 10 = 90   9. 840 ÷ 10 = 84   10. 1,080 ÷ 10 = 108   11. 290 ÷ 10 = 29   12. 550 ÷ 10 = 55   13. 1,180 ÷ 10 = 118   14. 170 ÷ 10 = 17   15. 10 ÷ 10 = 1   16. 370 ÷ 10 = 37   17. 90 ÷ 10 = 9   18. 1,720 ÷ 10 = 172   19. 780 ÷ 10 = 78   20. 990 ÷ 10 = 99   21. 0 ÷ 10 = 0   22. 280 ÷ 10 = 28

p. 127   2. 3 x 11 = 33   3. 1 x 11 = 11   4. 6 x 11 = 66   5. 11 x 11 = 121   6. 11 x 2 = 22   7. 11 x 5 = 55   8. 11 x 11 = 121   9. 8 x 11 = 88   10. 4 x 11 = 44   11. 11 x 11 = 121   12. 7 x 11 = 77   13. 10 x 11 = 110   14. 11 x 11 = 121   15. 11 x 0 = 0   16. 11 x 8 = 88   17. 11 x 9 = 99

p. 128   2. 44 ÷ 4 = 11   3. 99 ÷ 9 = 11   4. 88 ÷ 11 = 8   5. 121 ÷ 11 = 11   6. 11 ÷ 1 = 11   7. 77 ÷ 11 = 7   8. 121 ÷ 11 = 11   9. 110 ÷ 10 = 11   10. 11 ÷ 11 = 1   11. 121 ÷ 11 = 11   12. 66 ÷ 11 = 6   13. 0 ÷ 11 = 0   14. 55 ÷ 11 = 5   15. 121 ÷ 11 = 11   16. 99 ÷ 9 = 11   17. 121 ÷ 11 = 11

p. 134   2. 3 x 12 = 36   3. 1 x 12 = 12   4. 6 x 12 = 72   5. 12 x 11 = 132   6. 12 x 2 = 24   7. 12 x 5 = 60   8. 12 x 12 = 144   9. 8 x 12 = 96   10. 2 x 12 = 24   11. 9 x 12 = 108   12. 12 x 7 = 84   13. 10 x 12 = 120   14. 12 x 8 = 96   15. 12 x 0 = 0   16. 11 x 12 = 132   17. 5 x 12 = 60   18. 4 x 12 = 48   19. 9 x 12 = 108   20. 7 x 12 = 84   21. 12 x 3 = 36

p. 135   2. 48 ÷ 4 = 12   3. 108 ÷ 9 = 12   4. 48 ÷ 12 = 4   5. 96 ÷ 8 = 12   6. 12 ÷ 1 = 12   7. 84 ÷ 7 = 12   8. 120 ÷ 10 = 12   9. 24 ÷ 2 = 12   10. 72 ÷ 12 = 6   11. 24 ÷ 12 = 2   12. 12 ÷ 12 = 1   13. 132 ÷ 11 = 12   14. 84 ÷ 12 = 7   15. 60 ÷ 5 = 12   16. 132 ÷ 12 = 11   17. 60 ÷ 12 = 5   18. 0 ÷ 12 = 0   19. 96 ÷ 12 = 8   20. 72 ÷ 6 = 12   21. 36 ÷ 12 = 3

pp. 136–138   Visit **TheTimesMachine.com** for the answers to this big problem set.

Chapter 6

p. 144   2. 24   3. 15   4. 2   5. 50   6. 13   7. 1   8. 13   9. 8   10. 72   11. 36   12. 4   13. 36

p. 149   2. 12 x 5 = 60   3. 6 x 9 = 54   4. 6 x 12 = 72   5. 8 x 9 = 72   6. 10 x 11 = 110   7. 10 x 10 = 100   8. 7 x 8 = 56   9. 12 x 7 = 84   10. 6 x 11 = 66

p. 153   2. 140   3. 360   4. 450   5. 1,200   6. 1,600   7. 3,200   8. 250,000   9. 300,000   10. 6,300,000

Chapter 7

p. 158   2. 2 x 4 + 1 = 9 and 9 ÷ 2 = 4 R1   3. 3 x 4 + 1 = 13 and 13 ÷ 3 = 4 R1   4. 3 x 3 + 2 = 11 and 11 ÷ 3 = 3 R2   5. 4 x 2 + 2 = 10 and 10 ÷ 4 = 2 R2   6. 5 x 2 + 3 = 13 and 13 ÷ 5 = 2 R3   7. 6 x 3 + 1 = 19 and 19 ÷ 6 = 3 R1

p. 162   2. 4 R4   3. 6 R1   4. 6 R3   5. 4 R1   6. 9 R3   7. 11 R2   8. 7 R2   9. 7 R4   10. 6 R2   11. 4 R1   12. 10 R7   13. 12 R2   14. 8 R1   15. 2 R8   16. 8 R4   17. 12 R1

Chapter 8

p. 169   2. 138   3. 95   4. 252   5. 178   6. 182   7. 492   8. 704   9. 940   10. 1,236

p. 177   2. 128   3. 408   4. 539   5. 204   6. 198   7. 198   8. 690   9. 2,466   10. 615   11. 3,112   12. 412   13. 2,525

p. 183 2. **54 square centimeters, or 54 cm²** 3. **108 square meters, or 108 m²** 4. **200 square miles, or 200 mi²** 5. **49 square feet, or 49 ft²**

6. **21 square inches, or 21 in²** 7. **360 square millimeters, or 360 mm²**

Chapter 9

p. 189 **2.**

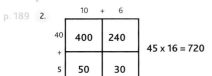

	10	+	6
40	400		240
+			
5	50		30

45 x 16 = 720

3.

	50	+	2
20	1000		40
+			
3	150		6

23 x 52 = 1,196

4.

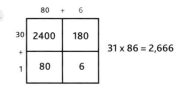

	80	+	6
30	2400		180
+			
1	80		6

31 x 86 = 2,666

5. **1,474** 6. **936** 7. **2,752** 8. **2,184** 9. **4,416** 10. **9,801**

p. 194 2. **621** 3. **1,581** 4. **2,666** 5. **342** 6. **1,768** 7. **864** 8. **2,860** 9. **1,474** 10. **1,012**

Chapter 10

p. 204 **2.**

369 ÷ 3 = 123

3.

848 ÷ 4 = 212

4.

308 ÷ 3 = 102 R2

2 left over!

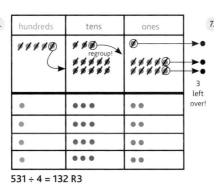

5.

255 ÷ 5 = 51

6.

531 ÷ 4 = 132 R3

7.

624 ÷ 6 = 104

3 left over!

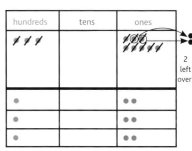

p. 214 2. **23** 3. **22** 4. **108** 5. **111** 6. **94 R3** 7. **175** 8. **157 R3** 9. **49** 10. **135** 11. **85 R7** 12. **437** 13. **333**

14. **107** 15. **48 R7** 16. **104 R1** 17. **1,091 R3** 18. **488 R6** 19. **657**

Be sure to check out **TheTimesMachine.com** for fully explained solutions to some of the trickier problems!

Index

"New Math" Translation Guide—for Grown-Ups!

Here's a guide to some newer math terms and methods, and where to find out more about them in the book. If you don't see what you're looking for, you can check the index, too!

AREA MODELS for two-digit multiplication (p. 184)	These are rectangles that show area! They can be used for two-digit multiplication by breaking up the problem into smaller problems. Then these answers are added together for the final answer!	13×15 $10 + 5$ 10 $10 \times 10 = 100$ $10 \times 5 = 50$ $+$ 3 $3 \times 10 = 30$ $3 \times 5 = 15$ $100 + 50 + 30 + 15 = \mathbf{195}$
ARRAYS (for multiplication, p. 17) (for division, p. 49)	These are the rectangles made up of dots (or other objects!) to represent simple multiplication (and division) problems.	This shows $3 \times 6 = 18$ ● ● ● ● ● ● ● ● ● ● ● ● ● ● ● ● ● ● and $18 \div 3 = 6$.
BOX (or Grid or Window) METHOD for multiplying two-digit numbers (p. 187)	Similar to the Area Models above, but with equal-size boxes, this method expands the factors to create four easier problems.	36×79: $70 + 9$ 30 2100 270 $+$ 6 420 54 $2100 + 270 + 420 + 54 = \mathbf{2,844}$
DECOMPOSING FACTORS (or DECOMPOSING MULTIPLICATION) (p. 41)	Another name for splitting big factors in a multiplication problem	Here, we'll split 12 into 2 and 10: $3 \times 12 =$ $3 \times 2 \ + \ 3 \times 10$

FACT FAMILY (p. 54)	This is a group of <u>multiplication and division facts</u> that all use the same (usually three) numbers. <u>Fact families</u> show that multiplication and division facts are "related" to each other.	Here's the full fact family for the numbers 7, 9, and 63: $7 \times 9 = 63 \quad 63 = 9 \times 7$ $9 \times 7 = 63 \quad 63 = 7 \times 9$ $63 \div 9 = 7 \quad 7 = 63 \div 9$ $63 \div 7 = 9 \quad 9 = 63 \div 7$
PARTIAL PRODUCTS (p. 171)	This is a popular way to teach multiplication without using regrouping (aka carrying). It involves writing separate partial products, and then adding them up to get the final answer.	$\begin{array}{r} 13 \\ \times\ 15 \\ \hline 15 \\ 50 \end{array}$ → $\begin{array}{r} 13 \\ \times\ 15 \\ \hline 15 \\ 50 \\ 30 \\ 100 \end{array}$ → Then add 'em up! $\begin{array}{r} 15 \\ 50 \\ 30 \\ +\ 100 \\ \hline 195 \end{array}$
PLACE VALUE CHART FOR DIVISION (p. 198)	This is a visual method of division using a chart and little disks to stand in for the numbers. Each disk can represent 1, 10, 100, etc., depending on what column it's in. See pp. 208–209 for a side-by-side comparison of this new method with good ol'-fashioned long division!	561, 3 bowls hundreds / tens / ones ● ● ● ● ● / ● ● ● ● ● ● / ●

Danica McKellar is the *New York Times* bestselling author of groundbreaking math books including *Do Not Open This Math Book,* picture books *Goodnight, Numbers* and *Ten Magic Butterflies,* and the middle school hit *Math Doesn't Suck.* She is a summa cum laude graduate of UCLA with a degree in mathematics and has published an original theorem in the prestigious *Journal of Physics* now known as the Chayes-McKellar-Winn theorem. She testified before Congress about encouraging women to excel in mathematics, has received many mathematics and educational awards, and was even named Person of the Week by ABC News for her unique and valuable work as a math education author.

Danica is also well known for her acting roles on *The Wonder Years, The West Wing,* multiple Hallmark Channel movies, and more. Visit **McKellarMath.com** to see the ever-growing line of Danica's helpful, fun math books for ages 0–16!

Don't miss these other McKellar Math books!

You can count on Danica to explain math concepts in fun, easy-to-digest ways
that give readers the tools they need to *succeed in math* and
build the confidence that comes from feeling smart.

Ages 0–5

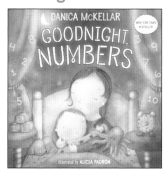

Ages 2–5

Ages 4–6

Ages 6–9

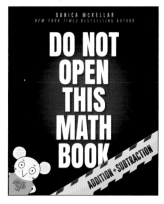

Ages 9–12

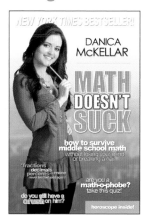

Visit **McKellarMath.com**

to see additional titles for teens.

Ages 11–13